The Next Benghazi

The Next Benghazi

Ezekiel Smith

BLACKSMITH 20 13 PUBLISHING

The Next Benghazi

by Ezekiel Smith

Copyright © 2022 Ezekiel Smith

ISBN 9781956904086

Printed in the United States of America

Published by Blacksmith LLC
Fayetteville, North Carolina

www.BlacksmithPublishing.com

Direct inquiries and/or orders to the above web address.

Contents

Acknowledgments

This book is dedicated to the victims of attacks that were predicted and could have been prevented. May we learn from the mistakes of the past and never repeat them.

Foreword

The bureaucratic world of government organizations is fickle and often counterproductive to national interests. It isn't that a favorable and productive organizational climate can't be created within government; it is that the human elements often obstruct it. Specific to government bureaucracies, senior management is made up of political appointees, which commonly are uninterested or willfully ignorant of the views of the rank-and-file career experts. Political appointments all too often correlate better with the amounts of money donated to political campaigns than bona fide qualifications. Further, they each come with an agenda and varying amounts of ego that in practice pit departments and agencies against each other even though they should be on the same team. Looking down the management ladder it isn't any better. Promotions and seniority for career employees are based primarily on time occupying a seat and the ability to "work the system," not merit and actual occupational competency. In fact, in government, many evaluation systems explicitly forbid using tangible and quantifiable metrics to evaluate personnel. The resulting effect creates a climate of mediocrity and poor management that fails to retain their best human talent.

It isn't just people that get suppressed by the bureaucracy, but ideas. The best ideas that make it to the top are stolen without recognition, ignored, or aggressively suppressed. Rare is it that a suggestion or initiative provided by someone outside of senior management gets any attention. Bureaucratic organizations reward employees that conform and never make waves. True initiative, independent thinking, and novel action are feared, not valued. Worse yet, bureaucracy enforces a class system that rigidly ensures employees "stay in their lane" and never challenge the senior dictates of their respective organization. Those that dare to speak out and do what's right are often punished and ostracized. However, they are not punished for being wrong. They are punished for drawing attention to the failures of the senior management of the bureaucracy. They are punished for not "waiting their turn." They are punished for not using the "proper channels." Combined, this bureaucratic atmosphere has a chilling effect on the necessity of an organization to be adaptive. In organizations charged with the responsibility to defend the United States of America, this organizational climate is not just detrimental, it has and it will continue to get Americans killed. Using a single word to describe this problem isn't hard. We call it "politics."

At approximately 9:40 PM, on September 11, 2012, a culmination of bureaucratic failures materialized as members of a terrorist organization known as Ansar al-Sharia attacked the US diplomatic compound in Benghazi, Libya. That attack resulted in the deaths of US Ambassador to Libya John Christopher Stevens and U.S. Foreign Service Information Management Officer Sean Smith. As the attack continued, the CIA Annex in Benghazi also fell under attack and Global Response Staff (GRS) employees, Tyrone Woods and Glen Doherty, subsequently lost their lives in the defense of the compound. This tragedy was preventable.

The sad reality was that anyone paying attention to the developments in Libya was not surprised by this attack. Appropriate actions should have been taken because of the widespread prior knowledge of the growing threat in Libya. The specific time and date of the attack may not have been known, but US Department of State employees readily acknowledged for months in advance that "it was only a matter of time." Further, in the months prior to September 11, 2012, numerous attacks were conducted against foreign missions in and around Benghazi proving the area was becoming increasingly hostile and dangerous. Employees coyly joked about who was heading to Libya next to "get killed by terrorists." Without any doubt, "everyone" knew that operations in Libya were going to get people killed, but it still happened. In fact, the view was so pervasive it did reach senior management but was still ignored. Specifically, on December 30, 2012, the United States Senate Committee on Homeland Security and Governmental Affairs released a damning report on the incompetence of senior management at the Department of State in the report, "Flashing Red: A Special Report on the Terrorist Attack at Benghazi." The report summarized a large amount of intelligence that assessed a major attack against Americans in Benghazi was likely, leadership knew about it, and action should have been taken. Specifically, the report cited in its findings:

Finding 1. In the months leading up to the attack on the Temporary Mission Facility in Benghazi, there was a large amount of evidence gathered by the U.S. Intelligence Community (IC) and from open sources that Benghazi was increasingly dangerous and unstable, and that a significant attack against American personnel there was becoming much more likely. **While this intelligence was effectively shared within the Intelligence Community (IC) and with key officials at the Department of State, it did not lead to a commensurate increase in security at Benghazi nor to a decision to close the American mission there, either of which would have been more than justified by the intelligence presented.**

The report went on stating:

The Committee has reviewed dozens of classified intelligence reports on the evolution of threats in Libya which were issued between February 2011 and September 11, 2012. We are precluded in this report from discussing the information in detail, but overall, ***these intelligence reports (as the ARB similarly noted) provide a clear and vivid picture of a rapidly deteriorating threat environment in eastern Libya—one that we believe should have been sufficient to inform policy-makers of the growing danger to U.S. facilities and personnel in that part of the country and the urgency of***

them doing something about it. *This information was effectively shared by the IC with key officials at the Department of State. For example, both the Deputy Assistant Secretary of State for International Programs in the Bureau of Diplomatic Security, Charlene Lamb, who was responsible for the security at more than 275 diplomatic facilities, and former Regional Security Officer (RSO) for Libya Eric Nordstrom, who was the principal security adviser to the U.S. Ambassador in Libya from September 21, 2011 to July 26, 2012, told the Committee that they had full access to all threat information from the IC about eastern Libya during the months before the attack of September 11, 2012.* **Yet the Department failed to take adequate action to protect its personnel there.**

Even Ambassador Stevens recognized this growing threat and journaled in his diary about the worrisome presence of terrorist groups such as al-Qaeda. Professional security officers in Libya were also acutely aware of the deficiencies in the US security posture and were sounding the alarm. Also very incriminating is the fact the RSO, Eric Nordstrom, requested additional security for the diplomatic mission in Benghazi twice from the State Department and was denied. The body of knowledge overwhelmingly supported the necessity of increased security measures. However, as events proved, no effective actions were ultimately taken by the State Department. This leaves many questions that have been inadequately answered or completely unresolved organizationally. However, the questions could be summed up by asking: "If everyone knew Libya was a ticking time bomb, why was nothing done to prevent it?"

The answer is politics. The internal organizational mechanisms to sound the alarm were used, but failed. This wasn't because leadership was not told of the threat as the post Benghazi investigation irrefutably demonstrated. The warning systems failed because the Department's leaders didn't listen and no one publicly challenged the group think at the top. This was not just a failure in leadership, but a lesson in political arrogance and government accountability. Had the Administration been publicly challenged on their lack of action, it is likely that the State Department would have been forced to improve security. Unfortunately, ignoring the reality of the threat fit the political agenda that strived to present US actions in Libya as a success during a contentious presidential election. No one wanted to acknowledge the problems because that would imply the incumbent Administration's policy was a disaster. Believing the lie was better than facing reality. Of course the facts on the ground told a very different story of Libya's post-revolution "progress," but nowhere were there "insiders" stepping up to speak externally. The failure of insiders to speak out should be startling considering they were the ones most likely to be killed in Libya. However, the compliant nature of those most intimately aware of these facts in Libya is a testament to the unwritten culture and pressure from within an organization to "shut up and color."

Ironically, in the Department's eyes, much had been done to improve the security situation in Libya. Some of these actions included increasing physical security, deploying additional Diplomatic Security (DS) Special Agents on a temporary duty (TDY) status, and tracking the situation in various operations (Ops) centers. "Bureaucratically," this checked the box and reduced the pressure to provide additional funding or security assets to the situation in Libya. The actions the Department had taken were seen as adequate based on a relative comparison of what the organization had done in the past. However, one could levy the indictment that senior leadership failed to adequately assess the fact that Libya was different from past situations and required a different approach. Had they done so, it is doubtful any lives would have been lost in Libya. Internally, the Department failed to predict new threats and take adequate security measures. As such, the security actions taken were deemed sufficient because they had worked in the past. In Libya, using a standard based on past precedent cost four Americans their lives. Noteworthy is the fact the average DS agent knew things were not good in Libya, but their concern seems to have either not have made it to senior management or failed to initiate action by senior management. As such, even if lines of communications did exist between the agents in the field and senior management, they did not work. Asking why agents felt Libya was a suicide mission was only one piece of the equation. Based on the information, management also needed to take action to mitigate the threats. Why this did not occur internally is worthy of study.

Now, almost a half-decade since the tragedy in Benghazi, has the State Department learned its lessons? The following rhetorical questions are illustrative of the answer:

Today, is it possible for one to navigate the bureaucratic roadblocks to get face time with senior department leadership? To those in power, is the seniority of the messenger more important than the content of the message? If someone had advance warning of a new threat or impending tragedy, who, if anyone, would listen? Is there an effective and reliable channel to get a message to senior leadership? Could a single person with threat information trigger a response by the State Department to mitigate the threat? Is sounding the alarm the type of action that is rewarded? If not, would taking additional steps to call attention to the inaction lead to being punished? Ultimately, is the Department now adequately addressing emerging threats and quickly institutionalizing mitigation strategies?

Dangerously, it appears that if the State Department had learned its lessons, it has now forgotten them. Today, a new threat has emerged that the Department is woefully unprepared to combat. If not effectively countered, the US will be dealing with another Benghazi moment in the near future. This isn't because the Department hasn't been warned. This is because the Department has willfully chosen not to take appropriate aggressive action. The proof of this is evident. To date, there is not a global program in effect to combat this threat and those that have tried have been effectively told to "stay in their lane." True, the Department is trying to play catch up and take "some" actions. However, like in 2012, the Department believes it is "taking action," but completely fails to see that their action falls short of what is needed. The Department is still failing to accurately predict future threats in a way the Department can and will take timely action. "After the fact," is not how a national security policy should be implemented. Further, the political climate is again becoming insular in order to protect its equities just as we witnessed in the run up to the election in 2012. As of now, internal access to the senior circle of leadership is firewalled whether they are aware of it or not. If the leadership doesn't open the door, listen, and act, we are only a short time from another Benghazi.

1

Red Skies in The Morning

The grounds of the United States Embassy in Tunisia were immaculate. Palm fronds slowly waived in the warm Mediterranean breeze casting dancing shadows over the closely clipped grass. The striking green stood out against the tan desert backdrop of Tunisia and was broken intermittently by bright flowers and bubbling fountains lined with blue tile mosaics. The juxtaposition of this oasis garden against the looming concrete walls topped by coiled barbed wire and cameras was found exclusively at US diplomatic facilities in high threat countries. This strange backdrop of an Eden in prison had become invisible to Ambassador Janet Reese as she hastily made her way along the tile walkway to the chancery. She had made this trip a thousand times during her tenure as the senior US diplomat in Tunisia. However, this morning Reese was late and the EAC (emergency action committee) was formed and waiting on her in the SCIF (Sensitive Compartmented Information Facility).

Recent activities of a violent Islamic terrorist organization known as Ansar al-Sharia, which operated mainly in Libya, had spurned a flurry of cables between Washington and Tunis. The terrorist organization had grown in strength since another terrorist organization known as the Islamic State or simply IS had been badly mauled in Syria by a combined military assault led by the US and Russia. The remaining IS members had dispersed from the battlefield and many retreated to the failed state of Libya where they reconstituted under Ansar-al-Sharia's organization. In Libya, the fighters helped Ansar al-Sharia factions overrun Libya's fractured government forces and firmly established themselves in the eastern half of the country. Since then, their attacks began to focus externally on disrupting Libya's neighbors in North Africa. Niger and Tunisia suffered the brunt of these attacks. Just in the last month, over a dozen major attacks occurred in Tunisia. The body count in Tunisia was becoming unprecedented. Over 124 tourists had been killed in a spate of hotel attacks that left its capital city of Tunis paralyzed and under martial law. Both the CIA and FBI had a growing sense from mounting intelligence that a major attack was being planned against the US Embassy in Tunis. Everyone knew it was only a matter of time before the terrorists' plan would go operational.

Ambassador Reese was dead set against increasing the threat level in Tunis much to the chagrin of the RSO (Regional Security Officer) Bill Burnett. Ambassador Reese was proud of her diplomatic work and had complete confidence that the assurances she received from the Tunisian Foreign Minister to protect the embassy were sufficient to mitigate any threat.

"Why the hell does Washington think that a war in Libya means Tunis should be on lockdown?" yelled Reese into her BlackBerry as she jumped out of her FAV (Fully Armored Vehicle) and walked towards Bill Burnett. Bill was waiting just outside of Post 1, which is the name for the main guard post at the entrance to the embassy. Post 1 was reliably manned by two Marines in pristine service Charlie uniforms identifiable by the classic short sleeve khaki shirt with ribbons, traditional olive-green wool slacks, and adorned with a shiny black gun belt sporting a Beretta M9 pistol. "Again, this is not new. We have been dealing with continual threats emanating from Libya since your beloved President Obama turned it over to the terrorists." Reese wasn't holding anything back as she continued to lecture her BlackBerry.

Bill already knew Reese was pissed. She never made her emotions much of a secret and could get away with it since she was unique in that she was a career diplomat that survived the political purges as a friend of the current Administration. Bill had previously mandated that her security detail modify their routes and departure times to the embassy as an increased security measure. However, Ambassador Reese was painfully punctual and the "unnecessary" detour today had delayed her arrival. Of course, she blamed the RSO. Why wouldn't she? The RSO was always the easy target for blame. After all, in her mind, all the guys over at Diplomatic Security (DS) seemed to do was find ways to limit her diplomatic outreach with their overcautious myopic focus on security. As far as Reese was concerned, DS had forgotten its mission was to enable diplomacy and instead was preventing it. "No, no, I am not ordering a voluntary departure. We have 49 children of diplomats here that I am not prepared to uproot in the middle of their school year." Reese snipped back to the unknown caller as she approached. Bill could tell this EAC was going to suck. Ambassador Reese was not in the mood to quibble with Washington over how to run her mission and she had already made up her mind, irrespective of the sensitive intelligence they had received overnight, the embassy was going to remain open for business.

Bill had already prepared his brief for the EAC after a sit down with the Chief of Station Will Chandler, which was the CIA's senior man in country. The intelligence was never this good and that had them both worried. The embassy was facing an imminent threat of some type of new style attack and had to close until the threat could be mitigated. Ambassador Reese was hell bent that wasn't going to occur and Washington had her back. This went directly to the White House so Bill knew he was fighting a battle already lost. The NSC (National Security Council), acting on behalf of the president, wasn't ready to close the doors on any diplomatic facility and admit it had failed to stop the spread of this terrorist organization into Tunisia. Major promises were made to the American public that this Administration was going to ruthlessly root out terrorists and the US wasn't about to run and hide. As the RSO, Bill was responsible for the security of the embassy and knew he was now in an impossible position.

This wasn't supposed to happen. As an RSO, Special Agent Bill Burnett was at the pinnacle of his career. Bill was a career Diplomatic Security Special Agent and was well respected. He had served in some of the worst locations globally and this assignment was supposed to be an R&R tour for him and his family. Bill repeatedly went through his checklists and asked for all the help he could get so that professionally, he had covered his ass should the worst occur. In fact, according to the Foreign Affairs Manual or "FAM," the security at the embassy met or exceeded all policy guidelines, which was a real unicorn as compared to most facilities that operated on a mountain of security waivers. The most visible portion of this was the Ambassador was now provided protection by Unit 1 from the Office of Mobile Security Deployments or simply MSD. MSD is the acronym for the US Department of State's elite tactical response team. For the last two weeks, MSD had been providing what they call a tactical support team or TST, which moved Ambassador Reese to and from her residence in an armored motorcade. The embassy also had increased host nation security. The Tunisian government was very helpful to the extent they could support. They provided two LAV V150 Commando Armored Personnel Carriers (APC) to protect the two main entrances to the embassy. They also provided an additional 20 soldiers per shift for the external perimeter. However, as was the norm for Tunisian soldiers, the main guns on the APC had no ammunition and the soldiers were not allowed to have loaded weapons or shoot without prior government approval. This meant that in effect, these forces were nothing more than show should a real attack occur. Nonetheless, the embassy had at least four layers of security that must be penetrated before anyone could gain physical access to the chancery and that was very unlikely. Anyone that did make it past the Tunisian military, the Tunisian police, the walls and the wire, and into the embassy grounds proper would have to face a highly trained and well-armed MSD team, a detachment or "Det" of bored and very aggressive Marine Security Guards (MSGs), a handful of JSOC (Joint Special Operations Command) operators and liaison agents from the FBI and DEA, and close to a dozen American trained armed local body guards before ever getting to the heavily armored chancery. Bill had also requested additional military support from the Marine Corps, but Ambassador Reese was not having her embassy turned into a military garrison and killed the plan. This seemed absurd since a Marine Expeditionary Unit (MEU) of some 1,100 infantry Marines were floating no more than 30 miles off the coast of Tunisia and willing to help. Nonetheless, Bill did manage to convince Ambassador Reese to request a Commander's In-extremis Force (CIF) unit comprised of soldiers from a rotating Special Forces Group be placed into standby to reinforce the embassy if it came under heavy attack. The CIF is specially-trained and equipped to conduct Direct Action (DA) and Counter Terrorism (CT) missions and is capable of rapid response. It wasn't perfect, but should ever a major attack come, the embassy could handle it.

2

The Invisible Assassin

Before the Ambassador had arrived, Bill had made sure to personally walk the embassy grounds. He was looking for anything that might suggest an attack was imminent. What disturbed him most wasn't that there were signs everywhere; it was that there were none. Today appeared like every other day, but perhaps, nicer. The temperature recently had been very hot and the last few days were much cooler and actually pleasurable for a summer in North Africa. People were out shopping again, kids were in school, traffic was flowing, the city was again bustling after the string of terrorist attacks, which had Tunis on lockdown for weeks. The local guard force, or LGF, was also light today. The recent lockdown of the city made it nearly impossible for all of the LGF to make it to work so nothing seemed out of place. Any other time, Bill would consider it a threat indicator if the LGF didn't show. However, today, the guard force commander had assured Bill the absence of seven guards was because they were still living in sectors of Tunis that were under military lockdown. As Bill walked back from the main entry control point (ECP), he made a mental note to have an assistant cross-check their stories later today.

"I am going to walk off compound to Café Mercato and grab an expresso before DC wakes up and my inbox explodes. You want to come Bill?" Pete Lowe asked. Pete was the embassy's General Services Officer. Pete could be a tough personality to work with, but nothing got done around the embassy without Pete's support so it was easier to play nice. Further, Pete wasn't that bad of a guy. He was just kind of awkward and about 100 pounds overweight. He had the personality where you could see how working from a back office on a compound gated off from the real world appealed to him. Nonetheless, Pete was a capable officer and kept the embassy, logistically speaking, running like a Swiss Watch. "Going to have to take a rain check Pete," Bill said. Bill didn't need to listen to the radio to know the motorcade was approaching. He could hear the heavy acceleration and braking of the vehicles as they raced through the gates. The MSD motorcade was moving at what Bill thought was a reckless speed, but he was willing to let them operate as they saw fit. MSD Unit 1 was made of top notch agents turned operators and they had operated in nearly every hot spot around the world. True, the operators were not the 20 something year old physical specimens that military special operations teams sported, but they more than made up for it with their experience and proficiency.

About a half of MSD was made up of former military and a good portion of them were in special operations units. If there was a common thread, it was they were all far brighter and arguably more refined than the average grunt. It was like the Foreign Service's very own gun club.

It was also the last enclave at the State Department where free thinking, frank speech, thick skin, and masculinity were still considered noble qualities. Respective of street credit with other tactical units, MSD operators on average are some of the best shooters of any agency, which is a bar by which all operators measure. Their training is set up to mirror many of the same schools and standards their fellow military units attend. This is critical for interoperability. However, MSD has a decidedly defensive mission to protect high level dignitaries, US personnel, and diplomatic facilities at high threat posts around the world. They are tasked with enabling diplomacy in the most dangerous places on Earth. In many ways, this makes their job much harder as they must work within more stringent mission limitations and constraints. Further, MSD often has to conduct missions in front of the media's cameras. Equally important is that MSD is known to play well with others. The average operator has 15 years or more of experience and has long since moved past having something to prove and knows the value of setting egos aside. In fact, the average age on most teams is probably around 34 years old putting their experience and judgment on par with old wise men relative to the average age of many military units. Add to the package that many can speak a second language, run their own businesses, and have advanced university degrees; the teams are very well rounded. The combined skills make MSD unique amongst special operations units, but they are the best in the world at what they do.

Drag racing across the manicured embassy grounds directly toward Post 1 was clearly a sport for the MSD drivers. At the last second the FAVs abruptly turned in unison forming a wedge of armored vehicles around the Ambassador's vehicle as they lurched to a halt. Before the Toyota Land Cruisers had even stopped a motley group of heavily armed MSD operators emerged from the FAVs and established a perimeter. Like a well-oiled machine, the operators said nothing as they took up positions covering 360 degrees around the protectee's vehicle in just seconds. Another second later, Ambassador Reese emerged surrounded by men that looked like they were either about to try out for a lumberjack competition or just got back from completing an Ironman Triathlon. Now physically surrounding Ambassador Reese, the detail quickly escorted her still yelling into her phone as she approached Post 1 where Bill was now waiting.

From the roof of the GSO warehouse located across the courtyard from the Chancery, Dave Nicholson and Matt Depenstein had a commanding view of the embassy compound. Dave and Matt were attached with MSD Unit 1 and acting as the unit's designated marksmen or "DM's" for this deployment. DM was the MSD's polite language for describing the guy that accurately puts bullets into the brain stem of bad guys from a long way away. However, the term DM was still too aggressive for the State Department so officially, the position was called a "DDM" or Designated Defensive Marksman. This lunacy pervaded all aspects of the State Department. Any other organization would just loosely use the term sniper, but not here! Dave was an excellent shot with his Mk 12 rifle and could rapidly nail head shots at 800 meters all day with the weapon system. The Mk 12 is custom designed for SEALs and other special operations units as a light weight, but highly accurate rifle capable of rapid, semi-automatic engagement. The rifle looked close enough to a standard M-16, but the Nightforce scope, Surefire suppressor, and AN/PEQ-15 aiming laser/illuminator made it immediately draw attention when he uncased it. It didn't take a military expert to know the man wielding that weapon was serious. "You bring your sunscreen today moron?" Matt ribbed. Dave was a full blown ginger and was already turning a bright crimson after just a short time being exposed to the North African sun. "Yes, your mom rubbed me down before I walked out of the house this morning." Dave quickly retorted. "I also got something better today Matt. This was sent all the way from the US, compliments of Amazon," Dave exclaimed in a tone of celebration. He then pulled what looked like a big tarp from a bag he had carried up to the roof. Without explaining what he had bought, Dave began setting up a sniper blind as a modified sun canopy. "I got this to match the color of the roof. I know it is still pretty overt, but I figure if I can pitch it low, it can't be any worse than just standing up here in the open with a spotting scope. Grab the corner and tie it to the air conditioning unit Matt." Dave said. Within about ten minutes, the pair managed to rig up the hide, which was remarkably effective. Unless you were perfectly level with the roof, it was very difficult to see them below the canopy, which was further camouflaged by tan netting they wrapped around the sides, which allowed air to flow in and them to see out, but others couldn't see in. Dave seemed satisfied with his accomplishment as he kicked back in is camp chair behind his rifle. "It's going to be a glorious day Matt." Matt's superstition didn't allow for him to hesitate and answered, "Don't fucking jinx us Dave."

"Did you hear their last radio traffic?" Dave asked Matt. "Yeah, they are five minutes out." Matt answered. Dave and Matt had been on the roof for over an hour scanning every nook and cranny looking for any sign of a threat. They were particularly concerned with an unfinished high rise building that overlooked the embassy compound from across the N9 highway. "Any changes?" Matt asked Dave. "Nothing." Came Dave's answer quickly. As the motorcade approached, both Matt and Dave found it easy to fixate on the armored vehicles moving through the checkpoints into the embassy compound, but they knew the threat wasn't coming from the motorcade. It would be from a window, a small hole, an alley, a rooftop, or some other place providing a degree of anonymity with line of sight access to the motorcade. Once the motorcade entered the internal compound, Dave audibly exhaled in relief. "Another package safely delivered." Dave prematurely stated. Matt didn't answer and was still staring through his spotting scope looking out at the unfinished high rise. "I got movement on 5 of white over at Bravo 2." Matt reported. The cryptic language was their internal verbal shorthand for the fifth floor facing their position of the building they had labeled B2. "They don't appear to be doing shit. No weapons, but they are watching the motorcade pretty intensely." Matt said. By this point, Dave was again behind his rifle staring at the small group of men watching the motorcade pull up to the chancery. "I don't see any weapons, but one guy definitely has some kind of iPad thing in his hand." Dave said. "Threat?" Matt asked. Dave, still behind his scope answered, "I can't tell, but unless watching Netflix is grounds for going hot, I think we are good." Going hot was a colloquial term for opening fire. Out of his peripheral, Matt could tell the Ambassador was about to get out and head into the chancery. Neither Matt nor Dave could quite quantify any threat, but they knew that if something was going to happen this morning, it would be now.

The morning sun was blinding and it was hard to see her. "Good morning Ambassador," Bill projected as Ambassador Reese approached hidden by the scrum of burly men toting an array of assault rifles and light machine guns. Still with her BlackBerry pressed to her head, Ambassador Reese simply nodded a curt acknowledgement as she gathered her purse and approached Bill. Bill noted a faint buzzing sound he couldn't place as the security detail approached, but in the rush, discounted it as a random industrial sound. Perhaps, Bill thought, the hum was an air conditioning unit kicking on or some power tools being used by the groundskeepers. Either way, it didn't register in anyone's head as a threat. Ambassador Reese only had a very short distance to cover. She literally had to walk the length of a two vehicles before she passed the armored doors of the chancery and entered sanctuary. Five steps short of Bill, a thunderous, metallic boom split the air. Ambassador Reese as well as two MSD operators appeared to instantly get slammed to the ground. "What the fuck was that?" Bill asked himself as he tried to pick himself up off the ground. No one responded, but he could hear muffled yells and see the remaining MSD team already moving the motionless Ambassador into Post 1 through a fog of smoke and dust.

From the roof, the flash caught the attention of Dave and Matt just as the sound of the blast hit them. The concussive force was noticeable, but mitigated by the distance. The blast was comparable to an 81mm mortar round detonation so the charge wasn't small, but it wasn't huge either. "What the hell just hit them?" Matt yelled. Dave was equally baffled and suggested, "Maybe a mortar." "No, nah, it can't be a mortar. At this range we would have heard the thud of it being launched. I didn't hear anything." Matt blurted as he processed out loud what they had just witnessed. "Did you see a launch? Maybe an RPG from B2?" Dave asked. "Again, no. I was watching B2 the whole time. There wasn't a damn launch of anything unless Hajj (Hajj and Hajji became derogatory slang loosely used by American soldiers in Iraq and Afghanistan to describe enemy fighters) now has suppressors for invisible RPGs." Matt continued. Dave's voice changed as he turned back to the motorcade. "You seeing what I am seeing Matt?" "Yeah Dave, yeah. Not looking good down there."

"Bravo Delta Actual, Bravo Delta Actual, Mongol One. Can you see what the hell is going on from your post?" John McEwan's voice echoed across the net. Bravo Delta was Dave's call sign. Mongol One was the call sign for the MSD motorcade element, which John was attached. "Negative Mongol One. Still assessing." Dave replied. "Roger, hold your pos and continue overwatch while we hardpoint and assess casualties." Mongol One directed. "Mongol One, where is the actual?" Dave asked. John was acting as the unit medic so it was atypical he would be on the net speaking for Mongol One "Actual," which was the team lead. Dave's gut fear wasn't long in being confirmed. "Bravo Delta, Actual is down. Just Holzrich and me. We have a lot of casualties we are assessing now." John's melancholy tone cut through the background noise as if amplified by speakers at a rock concert and conveyed everything Matt and Dave needed to know.

"Our day is about to get more sporting. We have vehicles approaching from 6 o'clock." Dave relayed in a remarkably relaxed manner considering the gravity of what just happened. Both Dave and Matt spun toward the oncoming vehicles now moving through the traffic circle just outside of the first Tunisian military check point about 800 meters from the embassy. "Can you tell if they are hostile?" Matt asked. "I can't tell. I can't see anyone in the damn vehicles." Dave answered in a confused tone. It was now 8:07 AM.

3

Institutional Blind Spots

Three months prior, Agent John McEwan sat at his desk near McLean, Virginia bitching about their management's inability to adapt. "Seriously, what the fuck is wrong with Myers? Does he not understand that we need to be ahead, not behind the tactical curve?" Ed Myers was currently the director of MSD and was a good guy, but very career minded and was terrified of making waves. In his defense, his hands were tied when it came to budget and ultimately the Countermeasures Directorate had final say over any operational deployment of a new system, which Myers had zero influence over. McEwan's teammate, James Holzrich, was used to McEwan's soap box diatribes and quickly piped up, "If you don't like it McEwan, why don't you stop bitching and do something about it yourself. "You know Jim; for once you have a productive point. I am going to email the director today." McEwan snapped back as Holzrich continued watching cat videos on YouTube without breaking his gaze. Dave Nicholson, sitting between the two, couldn't help but to add fuel to the fire while he perused gun porn on the internet. "You know McEwan shot the Triple Nickel standards yesterday clean and earned his coin Holzi. When are you going to get off your ass and show him up?"

McEwan cut them all off, "Seriously look at this video. It was uploaded yesterday by another group claiming allegiance to the Islamic State (IS). Have you seen anything like this before?"

The group assembled around McEwan's monitor. What played out was a crude version of a drone strike. The big problem was that it wasn't a US drone strike. It was a propaganda video showing an IS attack against a Kurdish Peshmerga checkpoint using what appeared to be a homemade drone. The small aircraft had been retrofitted to carry what looked like a small directional mine. You could see plastic explosives packed into a cavity that was then wrapped with what appeared to be ball bearings. The footage then switched to an aerial view of a dusty road, ostensibly in Iraq, as Middle Eastern music played in the background. Soon, what appeared to be three Toyota Hi-Lux pickups painted brown with machine guns mounted in their beds came into the frame. The trucks were parked behind big mounds of dirt and about ten soldiers could be seen milling about the checkpoint. The detail was amazingly good. You could tell the uniforms the soldiers were wearing were a Marine Corps digital camouflage pattern. At this point voices in the video began chanting "Allah-Akbar." From the footage, it appeared the drone then circled the military position and dove straight at it before the footage went blank. The video then picked back up with what appeared to be another drone's video of the attack. The footage showed the same military position. Then, for a brief second, you could see a small aircraft dart into the frame before it exploded showering the troops with shrapnel. The scene was repeated in slow motion. The drone then zoomed in on the carnage to show that half the soldiers were apparently dead and the others writhing in pain before the video abruptly cut off.

McEwan was the first to speak up. "I told you this was coming! How long before this information proliferates and we see it used to kill our guys? When is DS going to realize this is not an emerging threat, but here and now threat? We never do anything until after the fact. The Department always waits till someone gets killed before addressing the obvious, glaring threats." "Well, John, I will concede, you called this one before anyone else," Holzrich reluctantly admitted. "Yep, that definitely is going to be a problem down range and soon John," Dave said. Rounding out the opinions, Matt added his two cents. "Fucking scary if you ask me."

John began typing an email. "This is going to Countermeasures. I want to know where they are with this threat." "You are wasting your time McEwan," James interjected, "you already know they won't allocate a dime to something until after someone gets killed." "Well, at least I am going to do my part. If they aren't going to run a program, I will put the damn thing together myself. I have plenty of friends over at Fort Belvoir already working this issue." Looking over, James said; "No one gives a shit. You know that right?" Suddenly, Dave found reason for interest and got excited about the topic. "I heard the military has been doing some studies and found the most effective thing to take out these small drones is birds...like big ass hawks and eagles." Matt had maintained his bearing up until this point, but upon hearing the drone hunting bird story, he busted out laughing and threw a homemade bullshit flag he kept by his desk. "That's the most ridiculous plan yet. What, are we going to have dudes with falcons patrolling our perimeters?" Dave by this point had pulled up a half dozen YouTube videos of birds of prey hammering drones. "Yep. Watch and learn. Don't have to do anything. Just let them hunt. Bam! Did you see that Matt? Bad ass! Australian Eagle 1, DJI Phantom 0. How about those birds now Matt? This is pure 1980s Beastmaster shit and I am all about it. I am going to call up the assignments officer now and request 'falconer' for my next post in Kabul." With a single move, James then pushed back from his desk in his black office chair, spun, and came to a rolling stop an inch from John's ear. John continued typing while making great effort to act oblivious to James. Leaning forward in a manner to garner maximum theatric flourish, James whispered, placing a paused emphasis on each word: "waste...of...time."

4

Second Wave

"Come on." Yelled the medic from Unit 1 as he entered Post 1. Trying to focus through the thick, greenish ballistic glass of Post 1, which slightly distorted his view, Bill could see Ambassador Reese covered in blood. Her arm was nearly severed at the shoulder and her face was virtually unrecognizable. Where Ambassador Reese was holding a BlackBerry just seconds before was a dark cavity that appeared to be a gory mix of hair, skull, and brain matter. Her yellow pants suit was now crimson and the medic was vainly trying to stop arterial bleeds from at least two locations while it was clear he was wounded too. A growing red blotch had soaked through John McEwan's overly expensive and newly issued Crye Precision G-3 combat shirt. "Son-of-a-bitch. I just got this shirt too!" John thought. Blood was running out his cuff and off his fingers with the persistent drip-drip-drip of a faucet not completely turned off. He had sustained a ragged wound in his bicep from the blast. Examining the wound, John could see that shrapnel had cut a nasty gash across his arm, but it wasn't a game ender. He could still fight. He would dress it later with hemostatic gauze and a compression bandage. His face was also peppered with small bits of metal fragments that almost certainly would have blinded him if he had not been wearing wrap around eye protection. Yeah, the big Oakley's looked a little too 90s beach volleyballish compared to the contemporary style, but they were hard to beat for protection. Right now though, McEwan had bigger problems than a dated style.

Ambassador Reese had assumed a yellowish-blue skin tone with a waxy pale pallor and was completely unresponsive. McEwan knew she was circling the drain. Quickly reaching down he ripped open his IFAC (individual first aid kit) and pulled out an extra tourniquet and a package of hemostatic gauze. "Where the hell do I begin," McEwan thought to himself. Grabbing Reese's limp and noticeably cold arm, he began to scoop away congealed blood to find the severed artery. "Got it." McEwan felt the pulse of blood against his finger each time her heart beat. "Heart beat...good sign." McEwan continued to talk his way through the procedure as he did so many times during training using the "MARCH" medical mnemonic. The letters of "MARCH" stood for the order of treatment for a trauma patient. "M" was for stopping "M"assive hemorrhaging. Uncontrolled bleeding was now the first thing combat medics focused on because it was the first thing that would kill someone in seconds to minutes from an otherwise survivable injury. McEwan balled a piece of hemostatic gauze, which is designed to aid in clotting, and pressed it firmly against the Ambassador's severed artery. John was doing his best, but it was clear that he was attempting in vain to both hold pressure and pack the wound to her shoulder while trying to jam his knee in her thigh to stop the femoral bleed. "Damnit, she has lost too much blood! I can't even give her fluids to stabilize her blood pressure until I plug all the damn holes." McEwan wasn't talking to anyone in particular, but audibly venting his frustration somehow helped him maintain focus. "Ah shit...not good...not good." He could feel each beat of her heart getting fainter and fainter as the arteries spurting blood just seconds before were now just oozing. Then came the ominous death rattle a person never forgets. McEwan stopped working on the bleeds for a brief moment and tried to clear her airway again. A patent airway on the Ambassador was nearly impossible to maintain because of the damage to her head, face, and neck. Blood, teeth, and other various gore continued to block her airway. The best McEwan could do is prop her up so that her head drooped down allowing everything to drain out and away rather than back into her mouth and throat. Nonetheless, it was beyond even the best surgeon's capabilities to avert what now had become inevitable. Her wounds were too severe. Ambassador Reese gasped just a few more short spasmodic breaths and with a deep gurgle from her fluid filled lungs, exhaled for the last time.

Bill saw the MSD medic frantically working on Ambassador Reese. Gathering focus, he rolled over and started to get up and move toward her to help. However, before he could take a single step, a searing pain shot through his body, which paralyzed him in agony. Bill fell instantly back to the ground as the pain left him powerless and gasping for air. Looking down at the origin of the excruciating pain, Bill realized the calf of his left leg was now nothing more than a bloody mass of burned tissue and bone fragments. Still writhing in pain and unable to move, Bill felt a powerful hand grab him like a vise grip. The next thing he knew he was being dragged through the door into Post 1. The MSD operator that had just pulled him to cover then threw him a tourniquet. It was James Holzrich. Holzrich was lucky enough to have been on the opposite side of the armored vehicle when the blast occurred. James' uncanny luck had always served him well and brought him safely through the heydays of contracting in Baghdad with Black Water. Shielded from the shrapnel, he was one of the only operators not wounded at all. "Use it," the operator yelled as he moved on to prep for a counterattack. Bill tried to place the tourniquet around his thigh like he was trained, but the pain was too much to bear, and he nearly passed out with each attempt. The Marine guarding Post 1 saw him struggling so grabbed the tourniquet and finished the job. "Holy shit, that's painful." Bill gasped as the Marine quickly twisted the windlass pulling the thick black band of the tourniquet increasingly tighter with each turn. It felt like the corporal got it so damn tight his leg was being amputated, but the bleeding stopped. "Where is the Ambassador? Where is she? Is Ambassador Reese okay?" Bill uttered as he passed out looking at a bearded man standing over top of him covered in blood saying, "the Ambo is dead man."

Back on the roof, Dave and Matt watched as a dump truck was gathering speed and heading directly for the military checkpoint outside of the embassy as the "bump, bump, bump, bump, bump" of the heavy machine gun mounted on the armored vehicle could be heard. With each thud from the heavy gun, the armored vehicle violently shook and echoed a metallic ring. The dump truck still gaining momentum looked like something out of the movie Mad Max. It was painted black and had additional steel plates welded to the front to protect the engine and tires. Oddly, there was very little protecting the cab for the driver. This struck Matt as kind of odd. Usually, terrorists make great efforts to protect the driver so that he won't be shot before reaching his target. Thankfully, the Tunisian military was in no mood to mess with terrorists and had their defenses prepped and ready. They had paid a heavy price in the weeks prior for complacency. Now they were going to get their revenge and were all firing into the dump truck. However, as the guards at the checkpoint dumped surprisingly accurate fire into the truck now only fifty meters away from the serpentine obstacle made of concrete jersey barriers, it showed no signs of stopping. Belching black smoke as it shifted gears, the dump truck made its final turn into the barriers plowing barbed wire and other less consequential obstacles out of the way. Suddenly, the dump truck exploded in an enormous fireball that enveloped the entire checkpoint and sent shrapnel and debris flying thousands of feet. The massive concussive blast slammed Dave and Matt backwards literally knocking the wind out of them. Immediately after the blast, an eerie calm settled over the embassy compound. Everything was still and quiet except for the "sssshhhhh" of debris raining back down from the blast. Then the thuds, clangs, and dings of larger debris falling back to Earth began to pelt the embassy compound. Matt slowly lifted his head and could see Dave still burying his face into the roof gravel. Matt asked, "You okay Dave?" "What?" Dave replied. His ears were still ringing and although he could clearly see Matt's lips moving, he couldn't hear anything beyond what sounded like Charlie Brown's teacher. In any other situation, Dave would have considered this a blessing, but right now; they needed to figure out what the hell was going on. "Bravo Delta, Bravo Delta, you have eyes on the blast?" The radio suddenly crackled as Matt's voice started to become audible again.

"Roger, an SVBIED just took out the front gate and checkpoint." An SVBIED is the term for a vehicle borne improvised explosive device that was delivered by a driver that intentionally died in the blast to make sure the bomb got to its target. Thus, the "S" stands for suicide. "How bad is the damage?" "Not sure yet John, there is too much smoke, but it was massive. Stand-by." Matt and Dave began to assess the damage. The checkpoint was obliterated. The barrier plan was reduced to smoldering chunks of concrete and rebar scattered over an area the size of two football fields. The armored vehicle was destroyed and continued pouring black smoke and flames out of the gunner's hatch. There was nothing left of the dump truck. Only two charred corpses were visible. They were barely recognizable and heavily deformed, missing limbs, and appeared to have been turned inside out by the blast. Whatever was left of their skin and clothing was charred and still smoldering. Everyone else appeared to have been literally blown away. "John, I think the checkpoint is a total loss...I think everyone is dead," Dave stuttered into the radio without regard for proper radio etiquette and call signs, which were quickly going out the window as the situation deteriorated.

Meanwhile back inside the chancery, Bill Burnett woke up in excruciating pain. The MSGs and remaining MSD team had dragged Bill and Ambassador Reese's corpse deeper into Post 1 for triage and treatment with the other casualties. A Marine had placed a green wool "Army" blanket over the Ambassador's body. It was the best they could do for now. Bill could see other bodies also covered. As he surveyed the carnage, he shuttered. He could distinctly see a sky blue high heel sticking out from under another blanket he immediately recognized as the ambassador's OMS, which translated to "office management specialist." That is the State Department's fancy term for secretary. Irrespective of status or title, Bill could feel the nausea building. The OMS arrived only weeks ago fresh out of her training at the Foreign Service Institute (FSI) in Arlington, Virginia. She couldn't have been more than 22 years old and had a really bright future ahead of her. As the gravity of the losses started to sink in, Bill rolled to his side retching overwhelmed by guilt.

"This is a disaster," Bill thought. "How did they hit us?" The thought was tormenting him. Bill had taken all of the appropriate measures. He knew the embassy was locked down. "Maybe it was an insider? Yes, it had to been someone that smuggled a weapon into the compound. Maybe an RPG?" As he continued to try and put together the events of the last few minutes one disturbing clue didn't fit. "What was that buzzing before the blast?" Bill swore he remembered hearing a buzzing before the blast and it didn't fit with any weapons he had ever heard before. The searing pain returned and began to overwhelm Bill. It was so bad all he could think of was the pain. "Ahhh, fuck! I didn't know it was possible to feel this much pain," Bill yelled in agony. Any thought of piecing together the attack was now impossible. Bill's vision began to tunnel and fade as he could hear what sounded like the rushing of a train coming in his head. Bill slumped over unconscious once more.

"Need a good vein. Hold his arm...hold it out steady. Good." The post medical officer was administering IV fluids using a mix of Hextend and saline to quickly replace Bill's lost blood volume. "A few bags should be enough to stabilize Bill for the time being," the medical officer told the Marine helping treat the wounded. As Bill's fluid volume increased his blood pressure started to normalize and he regained consciousness. "Try to relax Bill," Wes the medical officer stated as he injected Ketamine into Bill for pain. Wes knew that as soon as Bill woke up he would be in teeth gnashing pain and was doing his best to get ahead of it. Almost immediately, the pain killer started to take effect. "Where in the hell do you hide this med doc? This shit is awesome." "Yes, it is and that is why it stays safely locked up in medical. If this doesn't work, I have some heavier stuff, but for now, you need to deal. Okay? OKAY!?" Wes was speaking directly to Bill's face, but Bill was already mentally working the problem set.

Feeling better as the pain killer took hold and his leg was re-bandaged with the latest style compression bandage to hit the market, Bill reached for his phone, which was remarkably still intact, and dialed the DS Command Center back in Virginia. The Command Center is Diplomatic Security's op center for the world. It is housed in a nondescript office building just outside of DC. They have direct lines to everyone important to include the White House. Further, they can pipe in live camera feeds from around the world and their interagency analysts standing by can access a trove of data in seconds. "You have about five minutes Bill before the drug takes full effect and you won't remember a damn thing so you better talk fast," Wes warned. "Noted." "DS Command Center, Agent Harris speaking. How may I help you?" "Shut up and listen closely. I only have one chance to convey this information. Are you ready?" "Umm, okay, yes," Agent Harris answered as he brought the call up for the rest of the command center to hear waving his arm to get the attention of the senior watch officer. "This is RSO Bill Burnett from the US Embassy in Tunisia. We are under attack. Ambassador Reese is dead and we have multiple casualties. Using chief of mission authority, I am requesting immediate military support. Did you get that?" Harris, after fielding many phone calls had his suspicions that this was a sick joke, but kept an even professional demeanor as he brought up the closed circuit television camera views from the embassy in Tunisia to verify the claims. "RSO Burnett, yes, I got your information. Let me work this....holy shit!" Agent Harris stood stunned and speechless for a moment as the live feed of the carnage came into full view on the panoramic screens paneling the command center's walls.

The feeds from the front gate were offline due to the damage from the SVBIED blast, but a more distant camera on the embassy's roof clearly showed the sprawling devastation. Smoke was billowing from all over the compound and a large portion of the outer perimeter wall had been destroyed. Further, the full color view from Post 1 was like a scene from the Texas Chainsaw Massacre. The structural damage was minimal, but there was blood everywhere. The feeds left zero doubt that the situation was dire and the embassy needed help now. The senior watch officer was already calling the Seventh Floor to update them on the situation while typing an alert to the White House and subordinate ops centers at DoD. The Seventh Floor is a reference for the main State Department building in Foggy Bottom where the Secretary of State's office resides. "Hello, Office of the Secretary of State, how may I direct your call?" "This is the senior watch officer at the DS Command Center. We have an emergency situation. I need to establish a secure line to the Secretary immediately. This is not a drill."

"Where are you? Are you okay, Bill?" Harris asked. "We are hard pointed at Post 1. My leg is pretty badly injured, but others are worse. We are going to need some significant medical support and a full scale mass casualty evacuation. As of now, I can't tell you if our perimeter is breached, but the chancery is secure right now." Harris answered promptly. "Okay Bill. Hang in there. We won't Benghazi you guys. It is going to take some time, but we are going to get you guys some help." At this point Bill began to drift out. MSD Operator James Holzrich picked up the phone as John assisted Wes treating casualties. "This is Agent Holzrich with the MSD unit supporting the embassy. What kind of support can you get us and how long is it going to take?"

5

Counterattack

"Here they come." Dave was behind his scope tracking about a dozen individuals dressed in black running towards the breach in the perimeter. "Threats?" Matt asked. Matt was still oriented in the opposite direction to try and cover all sides of the compound. A sharp bang of a suppressed Mk12 sufficed as Dave's answer. Although, in Hollywood flicks a "silencer" equates to an almost silent "tchu tchu" or "thwack" sound, in real life, a suppressor just reduces what still is a pretty loud crack of a supersonic round exiting the muzzle of the rifle. Matt turned his Leupold spotting scope in time to see the head of a male, probably in his early twenties, snap violently to the right as he fell to the ground dead. The American made Colt M4 he was carrying continued to slide across the ground for another few feet in front of his body before also coming to a rest. "Crack, crack, crack." Dave continued to pour accurate fire into the group of young jihadists charging the compound. "You got another guy about 100 yards to the left of the guy you just dropped. He is crouched behind a vehicle. See him?" Matt said. "Bang...fuck, missed. Bang. Bang." Dave continued to shoot as he adjusted his lead on a small guy carrying a pack and what looked like a remote control as he darted behind trees, buildings, and vehicles. He was further away and a lot faster than Dave had estimated and so it took a good three shots before he was leading him enough with his aim point before he finally found his mark. "He was a fast little fucker, but not fast enough." Dave stated as he communicated his shots with Matt. "I think we got all of them Matt. I am counting 11 Salafists that got their wish to meet Allah today." "Salafist? How the hell do you know these guys were Salafists Dave?" Matt asked. "I don't know Matt. When was the last time a bunch of Shia terrorists attacked us in a Sunni country Matt? Maybe Beirut 1983?" Matt pondered for a second. "Good point. Salafists they are. Oh, and for the record, I think Lebanon is like a 27% Shia and 27% Sunni." Dave just shook his head and stared back through his scope. "Nerd."

"Seriously, be a man and stop whining. Get a damn grip!" Mark Neise, another MSD operator known for his sensitive personality had already reached saturation with the growing crowd of helpless little sheep from the embassy. He was trying to hold security at the door and focus on whatever the hell just wasted half of the detail, but was being deafened by the shrill screaming and whaling of one specific diplomat named Caleb Stroh. On any normal day, this fine specimen of a Foreign Service Officer (FSO) had no problem telling you that he had graduated with honors from Dartmouth to make sure everyone knew how smart he was. He had gotten Mark's attention one afternoon in the embassy dining facility while telling a table of junior FSOs about how he felt sorry for the stupid guys in the military that had no other options in life. The FSO's arrogance alone was enough to piss off anyone, but the fact that today "he" was wearing an ostentatious outfit consisting of black rectangular rimmed glasses, a bright pink bowtie, a starched white shirt with suspenders, bright blue trousers, and a fancy pair of leather shoes without socks sealed the deal. Not to stereotype, but this guy was a Class A, pretentious douche bag. Somehow this "diplomat" managed to pass the Foreign Service Exam even though he had the diplomatic acumen of a bowl of oatmeal and had been giving diplomats a bad name ever since. In spite of being repulsed by Stroh's entire being, Mark stayed professional when it came to his protective duties. He no intention of letting the jihadists hurt any of the "sheeple" he was now protecting irrespective of Caleb's privileged and quite condescending attitude.

"So terribly rude and predictably common from people like you." Caleb sniveled with an overtly elitist air clearly intended to intimidate Mark. Mark just laughed, "I see why you didn't make the cut for Harvard." Although, Caleb had lived in an entitled bubble relative to the other 99% of the world's population, events had now burst that bubble and placed him squarely with everyone else in the "real world." This "real world" was where Mark, not Caleb, was the master and if Mr. Stroh wanted to live through this ordeal, he needed to learn is place. Perhaps in denial, perhaps because he had no other options, instead of continuing to engage Mark in verbal Judo, he retreated into the corner and curled up into the fetal position. John couldn't hear the exchange, but could guess the gist of what was said by reading the body language of those around Caleb. Even from John's distance, the glaring looks of contempt and disgust at Caleb were effortlessly appreciable. This was no more apparent than with the locally employed ladies working in the dining facility. Apparently, Tunisian women do know what real suffering and hardship looks like and didn't have any respect or patience for men like Caleb whining.

John McEwan turned from watching Mark at the door, gave one more look at the casualties, and began organizing the defense. He instinctively knew after nearly two decades of combat experience in nearly every war zone on Earth that the numbers of casualties would only grow if they didn't focus first on the fight. "Holzrich, grab the Gunny and get the MSG mustered." "Gunny" is the slang term for the Marine Corps rank of gunnery sergeant. Traditionally, the larger MSG Dets are run by a Gunny. Holzrich nodded and began rounding up the Marines now congregating at Post One and donning their additional combat gear stored in their react room.

"Hey. You guys need some additional shooters? Rick Sanchez, DEA; and this is Erik Smith, FBI. How can we help?" Rick and Erik were the liaison officers from their respective agencies coordinating operations in the region. They had been in their offices when the attack began just starting their second cup of coffee and getting down to the real work of the day. Initially, they had hard pointed in place when they heard the first blast, but once it was clear every gunslinger on compound was going to be needed, they jocked up with their tactical kits and made their way to Post 1. John had previously seen both men around the compound and had already sized them up as paramilitary. You might say they fit the look. They were both wearing Kuhl pants, a thick rigger's belt, Asolos hiking boots, and plaid Columbia short sleeve shirt. Hell, it might as well have been an issued uniform and made them very easy to pick them out. However, it was the angular profile of the muzzle of their Glock 19 pistols protruding below their untucked cover shirts that made it obvious. One would think any professional would be diligent in concealing a firearm, but that would be in the normal world. Embassy life was different. The Glock, which was carried by everyone from GRS and CIA case officers to federal agents, conveyed a certain degree of mystique and status to the bearer. As such, it was a piece of coveted flare for folks seeking attention while pretending to be covert and yes, it can definitely confirm major douche bag status. Nonetheless, John needed all the help he could get and both DEA attachés and FBI Hostage Rescue Team (HRT) agents he had previously worked with were solid professionals. Sizing them up, John offered a handshake saying, "John McEwan. Good to see you guys. We need to sweep the compound grounds and try to get to the main gate. We are going to need every person we can get. No doubt this thing isn't over and my guess is that they will be coming through that breech sooner rather than later. There were a lot of folks out there too. We have wounded that need help." "You just say when John and we are with you," Rick said. The men shook hands and made their way to link up with the rest of the group.

Holzrich returned from the ready room with eight Marines. "Where are the rest Jim?" John asked. The Gunny spoke up. "Sir, this is it. The rest of the Det is in the Marine house hard pointed. They did the shift change at 0700 and were about to hit the rack. Now they are holding their position waiting for orders." Looking up after quickly checking his Mk18, John issued a quick order. "Okay Gunny, tell them to be ready to conduct a link up with us in ten minutes. We will be moving from Post 1 to the Marine house. Make sure they are ready to let us in and that they don't shoot us." Turning to the rest, "Jim, Erik, Rick, you guys move with me. Let's get some guns by the door and clear the exit. Once you guys have the exit cleared; hold it. Then when you are ready, give the nod and I will push the first four Marines and Gunny across the courtyard."

Turning back to Wes and a growing group of embassy personnel, he looked down at Bill Burnett. "Wes, when Bill comes around, give him the update. We don't have time to clear the building right now. As you know, we got bigger problems outside that need attending." Wes nodded and went back to treating the casualties. As John walked back to the Marines, which were prepping for the defense, he noticed the post's econ officer, Jennifer Stephens. Jen was in an earlier medical class he had given to embassy personnel and he had remembered her to be a sharp young women...not bad looking either. "Hey Jen, I need your help. We can't clear the building right now, but can you work on getting a good head count of who is here? If nothing else, just help Wes treat the wounded and try to coral the folks in this central area until we can get back. Remember MARCH from class right? Good-to-go?" Jen was still in shock, but the focus on an actual task helped steady her nerves. Her training actually started to kick in. Previous to deploying to post last year, she had attended a course back in West Virginia the students called "crash bang." It was a short course designed to prepare US personnel to serve overseas in high threat areas. It was by no means an operator level course and is generally dismissed as a joke, but it was designed to give the typical "civilian" the tools to survive just this type of situation. Taking a deep breath, Jen gathered herself and looked John directly in the eyes. "Yes, uhh...yes, I can do that John." Still assessing her penetrating eye contact, John tossed here a duffle bag filled with medical supplies to break up what was quickly becoming a slightly awkward moment. Thankfully, the embassy had the foresight and good sense to stage medical gear and equipment in designated areas for these types of contingencies. "I know you can Jen. Thanks." John said. Digging into the duffle bag, she pulled out a box of compression dressings and packages of Celox hemostatic gauze. As Jen carried the medical supplies toward the casualty collection point, she looked back at John to reassure him. "We can take care of things in here John; go do what you need to do." Jen held her gaze on John longer than she needed as the remaining agents headed out, but John was oblivious. He was focused on clearing the compound.

"Clear!" Jim was now controlling the exit with Rick and Erik and letting the composite team know it was fight time. Firmly gripping the Marine's shoulder at the front of the stack, John signaled for them to go as he yelled back, "Moving!" Instantly, the first four Marines sprinted into action as they raced across the open court yard to positions of cover on the far side. "Ah shit!" Yelled Jim. "What?" John quickly asked with concern. "No worries McEwan...I thought one of the Marines got hit. He just tripped and busted his rear end. He's good." Jim reported. Still holding his breath, John slowly exhaled. "Good to know Jim." "Set!" The Marine fireteam leader yelled back to signal they were ready for the next group to move. "Moving!" John yelled as he gave a smack on the back of the last four Marines and the Gunny to head across. "There's always one," Jim said as a Lance Corporal sprinted across the courtyard leaving what we called a "yard sale" of gear that fell out or was dropped. "Canteen, magazine, random item, MRE,..." Jim continued to name off each item as it fell. "Seriously, it is a really good thing that Marine's posterior is firmly attached to his body or he would lose that too." Jim continued. "Set" Yelled the Gunny. "Okay, gents, let's move," John said. Rick and Erik and the remaining MSD operators sprinted out of the building letting the heavy armored entry door swing shut and lock behind them.

"Coming in!" Gunny yelled as he got to the reinforced door sealing off the Marine House. "Click," the maglock on the door yielded to him allowing it to be opened. Within seconds the entire team poured through the door and into the house. John's eyes took a second to adjust from the bright desert sun to the dimly lit room kept dark and cool from blackout blinds covering all of the windows. "Oh boy," John thought as he saw eight Marines in various stages of dress and kit. His favorite was the Marine wearing a helmet and body armor over a pair of green yum-yums (super short silky running shorts) and a pair of black flip-flops. "They still issue those shorts?" John thought to himself before snapping back to the work at hand. John called to the Marines. "We will hold here for a few. Fix yourselves. As soon as you are up, we are heading out to start clearing the compound grounds from here to the front gate. Any casualties will be moved by the rear fireteam back to the last rally point. If we get contact, move to cover and eliminate the threat." "Mongol One, this is Bravo Delta." John keyed his mic, "Go ahead Bravo Delta." Dave looked at Matt as he again keyed his mic. "Mongol One, we have you visual. Just give us the heads up before you move. We have overwatch." Knowing Dave and Matt were providing covering fires gave John comfort they weren't flying blind. "Roger Bravo Delta, out." "Sir, the Marines are up. We are ready to move." Gunny reported. John took a drink from his Camelbak, "Okay gents, stack up on the door. On my signal, we will move in fireteams to the GSO compound. As soon as we get the second team up there, breach the door, and clear into the building." Holzrich seeing John busy, keyed his mic, "Bravo Delta, this is Mongol One. Are we clear to move?" Dave and Matt scanned their sectors. Everything was still and quite after the second wave. "Mongol One, you are cleared to move." Jim looked at John and gave him the nod. John turned to the men now staged. "Go!"

Amped up, instead of moving in units of four or five, the first two fireteams sprinted en masse out of the door. Racing up to the door of the GSO warehouse, the Marine with the key was preempted when the door swung open. As the door opened, no one was hesitating and the Marines poured through the entrance flattening a terrified Pete Lowe in the process. Pete probably wasn't in real danger of getting shot, but a smart man would have at least stepped to the side before surprising a bunch of armed Marines raging on adrenaline. By the time the last group got into the building, Pete had managed to struggle back to his feet. "Damn Pete, you okay Mark asked." Dusting himself off, Pete looked back at armed men standing around him, "Well, I was doing well until I was trampled by a stampede of Marines. Ha...don't take it the wrong way guys. I am really glad to see you. I just didn't think I was going to get gut punched and knocked to the ground when I opened the door." "Yeah, sorry about that. We didn't know if you were hostile," a Marine admitted. John took Pete by the shoulder as the rest of the fireteams fanned out through the building. "Is there anyone else in the building Pete?" Pete thought for a second, "Yes, I left five or six local staff in the office at the end of the hall." "Okay, let's get everyone staged here." John said. "Holzi, take the team and clear to the end of the hall. There should be five or six LES (locally employed staff) holed up at the end. Grab them, clear them, and then get them back here so we can move." Jim, Mark, Erik, and Rick moved quickly down the corridor clearing on the move. As they cleared each room, Mark took a fat Sharpie and drew a big check mark on the door so others knew it had been cleared. Within minutes they were back with six terrified locals. Gunny also reported back that the rest of the building was clear. "John, what do you want to do with them?" Mark asked. "Hmm...good question. Hold tight for one Mark." John said as he called up to Dave to get an updated. "Bravo Delta, how are we looking?" Dave grabbed his mic, "Still good brother." John turned to the Marines. "Gunny, I want you to send one fireteam back with the LES to the chancery. I want them to hold security there and be prepared to act as a QRF. Cool?" "No problem sir. Corporal, you heard the man. Take your fireteam and the LES's back to the chancery and be prepared to bail us out." Gunny ordered. "Yes sir," was all that was said as the corporal moved his fireteam with Pete and the LES back to the chancery. Aside from Pete nearly having a heart attack from running 50 yards, it went remarkably smoothly. John watched intently as he provided cover until the doors maglocked behind the last guy into the chancery.

"Bravo Delta, are you able to see any wounded?" John called over the radio. "Mongol One, best guess is move out and check the area around the gym. Nothing left beyond that point. The entire ECP is destroyed." Dave called back. "Roger, we are going to leave one fireteam here to hold the ground floor for you while we move to the gym. Out." John replied. Turning to the remaining composite team, John gave another quick order. "Gunny, I need you to push your two fireteams to the corners of the gym and hold security for the outer cordon. Once you guys are set, we will breach and clear the buildings." "Roger that sir." Gunny replied. "Okay, let's move," John really didn't need to say a word. The team was already falling into a rhythm and keying off of his movements.

Both fireteams sprinted from the door way, moving in bounds to covered positions. Bounding was a military term for what amounted to combat leap frog. In its most basic form, while one person was behind some cover providing overwatch, the other person would sprint to the next piece of cover. Once set up, the roles switched and the other person sprinted forward. In this way they teams continued to move forward until they reached their objectives on the opposite corners of the gym complex facing the now obliterated ECP. Seeing the Marines in position, John and the rest of the entry team moved out. They weren't sprinting, but rather, moving in a very controlled fashion toward their designated breach point. Each man covered a sector and moved at a speed he could immediately engage a target should it present. Dave and Matt were also scanning some distance in front of them. As they approached the door, James Holzrich, holding a modified Halligan Tool moved forward to make the breach. Reaching the door's threshold, James gently tested the knob as Mark provided cover. Feeling no play, the door was locked as expected. James braced himself and thrust the tool into the doorjamb adjacent to the locking mechanism. With one sharp jerk, the door frame splintered and the door swung open. James gave it a slight kick to make sure it swung fully clear. At the same time, Mark yelled, "bang out," as he pitched a flash bang into the room. A flash bang is a type of hand grenade that is used as a distraction device and is sometimes called a stun grenade. It isn't just a catchy name. The device actually stuns and disorients occupants in a room with its bright flash and loud bang, but doesn't otherwise hurt anyone. MSD preferred these less than lethal devices because their mission set typically called for operating in close proximity to noncombatants with a zero bar for collateral damage (i.e. killing a good guy). This was a far tougher mission than just killing everyone. Nonetheless, such was their mission.

A second after the bang went into the room, there was a loud "boom" as the device detonated and the entire team flowed instantly into the room clearing their sectors. Once their foothold was established, the teams changed tactics. Rather than a high risk dynamic style entry used to breach into the building, they fell into a much more controlled and methodical clearing technique. The men moved smoothly, but quickly down the halls. As they approached a door, the lead man posted short of the door and aimed in at the threshold as the second man flowed past, scanning the doorway from the outside. The intent was to use angles to clear as much of the room from the outside before ever entering. If there was a threat, it could be immediately taken out before committing an entire team to the room. If nothing was observed in the scans, the team set up on the door from opposite and interlocking angles, the door was breached, and entry was made after any threats were neutralized. For open doors, once the second man scanned across the doorway, the team entered focusing on the small portions of the room left uncleared, which generally were just corners and behind obstacles like desks and bookcases. At these ranges, John didn't need to "aim." He could instinctively point shoot head shots all day by simply shouldering his weapon and looking over his sites at the target as he fired. This allowed him to maintain maximum situational awareness, but quickly engage targets.

The team continued their methodical clearing room after room finding nothing. "Dry hole. Coming out." Mark yelled as he cleared into an empty hall closet. As the team turned to clear the last section of the building, Rick Sanchez, froze holding his weapon at the ready position in his shoulder as he aimed in on something at the end of the hallway. This was a universal signal to the other operators he had seen something not right or a bad guy in his sights. Keying the pressure switch on his rifle, Rick's Surefire X300 beamed 600 lumens of blinding light illuminating blood smears that led into the last room. Holding on the doorway, the rest of the team cleared their way to the edge of the door and then rapidly made entry. Holzrich and Neise were the first two inside the room. As John approached the door's threshold, he already heard the call. "Medic!" Flowing in, John still cleared his sector and then turned to James kneeling beside a wounded man wearing a guard's uniform. He was injured pretty badly, but was alive. As the guard tried to mouth some words he doubled over gasping for breath. John quickly moved up beside the man swinging his M9 Medical pack off his back as he tapped James on the back signaling he would take it from here. James and the rest of the team then set up security as John began to assess the casualty. Most of the bleeding was superficial, but as John raked his hands over the guard's chest looking for severe injuries, he felt his finger slide into a cavity that definitely wasn't supposed to be there. The guard had taken either shrapnel or a gunshot wound to the chest, which was quickly developing into a serious tension pneumothorax, which was crushing his lung against his heart, arteries, and veins preventing blood return. This was a textbook case. John was looking for a hole and worsening breathing as the key clinical signs, but the guard also showed classic tracheal deviation and neck vein distention, which are signs of the condition in a severe, latent stage before death. Reaching into his pack, John pulled out a large bore 14 gauge needle, felt for what he guessed to be the 2nd intercostal space in the midclavicular line of the chest, and then in a motion that made anyone watching wince, slid the needle slowly into his guard's chest until an audible hissing sound of air escaping could be heard. The guard's eyes nearly popped out as John "stabbed" him with the giant needle in the chest. Although, one would think John was killing him, he was most definitely saving the man's life. The language barrier prevented John from explaining what he was about to do so he just did it before the man could react. Almost immediately, the guard gasped in air as if he has suddenly been resurrected by the hand of God. However, there was nothing magical about the procedure. John simply released the pressure that had built up inside the man's chest. This procedure would probably need to

45

be repeated until he received definitive care, but for now, he was able to function and would live. Checking then for an exit wound, John quickly applied a HyFin Chest Seal, which is essentially a piece of plastic coated on one side with a super sticky adhesive, over the hole. After John sealed up the man's chest, he called for Gunny to send up the fireteam they had sent back to the chancery as a QRF to transport the man to the casualty collection point for additional treatment and care.

As they held their position waiting for the QRF to take the guard back to the chancery, with a look of terror, the man began mumbling something about being attacked by ghosts. Mark just began to laugh and said, "Yeah, they are all ghosts now." The man seemed to understand Mark and didn't think it was funny. Instead, he got more animated and adamant that they were attacked by ghosts. Peaking John's interest, he knelt back down beside the man as he packed up his med bag and said, "What ghosts?" The man looked at John and said, "The truck...the truck. It had no driver." As the rest of the team busted out laughing, John stopped everything he was doing and looked back at the guard. "What did you say?" The guard repeated his prior statement, "The truck had no driver." The gravity of the situation hit John. For the first time today, John had a clue about what was going on and how the terrorists managed to hit them as they dropped the ambassador. Keying his mic, John in a very serious tone called Dave and Matt. "Bravo Delta, quick question. Anyone you dropped out there carrying any odd or atypical gear?" Matt looked at Dave. "Where do you think he is going with this Dave?" "Don't know Matt, maybe suicide vests or something." Dave said. "What about that dude with the pack you dropped by the car? Think he was carrying a breaching charge or something" Matt said. Dave picked up the radio. "Mongol One, not tracking anything special, but we dropped a guy about 100 yards from your position in the car lot carrying a pack. Is that odd enough for you?" John thought for a second before replying. "Yes, that will work. I need to go check out the bag. Can you provide cover?" Dave quickly answered, "Roger Mongol One, but be careful, don't blow yourself up." John wasn't worried about a bomb in the pack. In fact, if he was right, it would be much worse.

"Gents, I need you to cover me as I move out to check the body in the car lot. I think I know what the hell is going on, but need some confirmation." John announced to the team. Nodding, Gunny just said, "No problem, the Marines will hold their positions." Moving back through the building, the team took no chances and cleared on the move. They didn't have enough people to clear and hold so as soon as they left an area, they had to treat it as unsecure. Moving past the racks of weights in the gym, they set up by the emergency exit. The exit provided access to the shortest route from the building to the corpse lying between two worn out black Chevy Suburbans that someone had mentioned were from when the US evacuated from their embassy in Libya years prior. That didn't matter now. There were about 75 yards of open field with short grass and no cover before reaching a low cinderblock wall and the parking lot. "Bravo Delta, we are in position. Do you have any movement?" John radioed. Dave was scanning the buildings across the street intently, he had thought he had seen movement, but couldn't find anything now. Matt radioed back, "You are clear out." John dropped his med back and any additional gear by the door. This was going to be a dead sprint. He knew that if there was a sniper across the street hiding in one of the buildings, he didn't have a chance of returning fire. Speed and surprise were going to have to work.

Taking one last deep breath John nodded. James threw open the door and John sprinted across the field. The 75 yards were a blur of short, dried grass strewn with trash bags as if they were Christmas tree ornaments. Fifty yards into it, John was not feeling as though this was moving by quickly. In fact, it seemed like a damn eternity. He was breathing deeply as he tried to pump his legs faster, but he was running out of gas. The weight of his kit in the now hot day coupled with the crushing of the armor plates on his chest made him feel as if he couldn't breathe. He had trained under these conditions relentlessly, but it somehow always falls short of the real thing. Staying focused on the wall; John stuck his vault perfectly, but blew the landing. Planting his hands on the top of the wall, he launched himself up and sideways over the wall in one smooth acrobatic motion. However, the ground was a lot further than he had estimated. With the finesse of a wounded grizzly, John rolled off the backside unable to stop his momentum and hit the sun baked ground with a thud. "Ahh, that hurt." John blurted out as he squinted staring up into the North African sun now overhead. "That was looking good until the end. You good?" Dave radioed as he watched the gymnastic train wreck through his scope. John, still laying there sprawled out like the Vitruvian Man moved only his arm to key his mic. "Yep, I am good. However, my pride isn't looking good. You are welcome to leave that out of the after action report." As John sat up he could feel the sweat beads stinging his eyes as he surveyed the scene taking a moment to catch his breath. Being careful to stay low, John crawled toward the location of the dead terrorist. "Bad idea," John thought as the heat from the hard packed desert ground began to burn the less protected areas of his body. By less protected he meant his groin. "No. Horrible idea." John continued to talk to himself as he moved forward. Moving to a shaded area under the back bumper of a tan Suburban, John scanned the area again and saw the black outline of the body about a car length away. He took another look around and cautiously crawled up to the body, which was already bloating and drawing a swarm of giant flies that had an iridescent greenish color. As he got to the body, he quickly checked for anything the resembled wires or explosives indicative of a suicide vest or bomb. Seeing none, he took a deep breath and gently pulled the pack from the body as he cringed in expectation of something blowing up as if that would somehow shield him from the blast. As the stiff body rolled free of the pack, John tried in vain to ignore the face of the dead terrorist staring at him who was noticeably missing the back side of his skull and spattered with bloody grey brain matter. Once the pack was free of the corpse, John drug it back to cover behind a worn out armored Land Cruiser. Sitting up against the front wheel, John slowly unzipped the top of the pack. As he opened the pack, John's fears were immediately confirmed.

Matt was watching John's movements inside the parking lot when he saw him suddenly jump up and begin running like hell back to the gym. "Dave, I think your right about that bomb. John is getting out of there as fast as he can." Matt relayed to Dave. "That sandbagging hillbilly. He sure as hell never ran that fast on a PT test." Matt continued. Dave was still studying the building still under construction he thought he had previously seen movement in. He could now see some young men messing with something in a back room, but couldn't tell what they were doing. All he knew is he couldn't see any weapons and the 10x magnification on his scope wasn't cutting it. Dave, still on his scope, yelled back to Matt, who was busy narrating John's sprint back across no man's land. "Hey...need you to focus over here for a minute. Crank the power up on the spotting scope and see if you can tell what the hell these guys are doing. Black side of the third floor. Check'em out." Matt loosened his tripod and swung his scope to the building as he acquired the men. "Can you see them? Three males about to walk behind the wall." Dave asked. "I've got nothing man. Missed them." Matt answered.

Erik and Rick were still holding the back exit area of the gym when they saw John leap over the wall in a full sprint back to their position. A moment later, John burst through the doorway into the gym that ended in a slide as he sprawled out once more. "Okay, this is getting stupid now. I really got to work on my landings," John thought. Laying there on his back again keying his mic, John managed between gasping for air to say, "Mongol One to all units. Get off the rooftops and get inside. I repeat...get off the damn roof and get inside right now. They are using drones. They also have some Elon Musk self-driving car shit going on too." Rolling to his side, John yelled to Gunny, "Pull back your Marines to the GSO warehouse right now. Get them inside." Mark looked at John as if he had lost his damn mind. "McEwan, what the hell is going on?" John worked up now to a knee and looked at the team, "Drones. They are using drones. The whole damn attack was launched using remotely controlled aircraft and vehicles rigged with explosives. The jihadi Dave shot in the parking lot was carrying a pack full of drones and a remote. They are picking us off from standoff positions and this isn't over." Mark didn't seem to grasp the threat posed by an enemy armed with a fleet of weaponized drones. "Easy day man. Shoot'em down." John looked back over at Mark in astonishment. "Are you serious? Like you shot the drone down that took out Ambassador Reese and our team leader?"

Matt and Dave heard the order over the radio and started to pack up to displace to a new position. Dave, cocking his head, stopped as things began to make sense. "Matt, that's what the hell those guys were doing when the motorcade was hit. They had a damn remote and I missed it as a threat. I can't believe I am saying this, but McEwan

51

was right about all the UAV threat stuff. It's legit. We missed it man." Just as Dave was about to grab his pack, he saw the movement again in the building. "The bastards are back Matt." Dave said as he dropped back behind his rifle. Matt looked over and could see the movement just as he heard the same ominous buzzing sound overhead that preceded the motorcade attack. "Come on Dave, we have to get out of here right now...right now damnit! I hear a drone overhead." "No, screw that Matt. I am nailing these bastards this time." Dave answered.

Unknown to Matt and Dave, approximately 200 feet directly above their position, another modified, commercial off the shelf (COTS) drone was flying toward the embassy carrying an 82mm mortar round with an activated point detonating fuse. All the drone needed to do was drop the mortar round and it would kill anything within a roughly 35 meter radius. Across the street, three jihadists stared at an iPad screen plugged into an RC controller showing a crystal clear, high definition camera view of two Americans exposed on the roof of a building. The terrorists had proved to be far more tech savvy than the State Department had assessed or even imagined. As the saying goes, necessity is the mother of invention and in this case, attack after attack against hardened American facilities had failed. The terrorists needed a new tactic and drones were the answer. Not only did unmanned systems provide the ability to bypass all of the physical security barriers, but they allowed the terrorists to press an attack with impunity. In fact, this was the first time anyone had attempted to take down an entire compound using drones almost exclusively. To support the operation, the terrorist cell had managed to build a rudimentary, but functional targeting app for aerial drones that could drop a mortar round from 400 feet in the air on a vehicle size target and hit it about 90% of the time if it was stationary. This was no small feat and the proof of the devastation was being broadcast live across huge television monitors in operations centers now across the US Government. The coordinated attacks across the city had been incredibly effective by any measure and every news agency in the world was now covering what they had dubbed the "Siege of Tunis." Although they lost some fighters by mistakenly committing them to the fight too soon, they had paralyzed the city, pinned down the Tunisian military, and were quickly preparing to finish off the US Embassy, which was their primary target from the start. The plan was going incredibly well and they were about to kill two more of the Crusader's Special Forces on the roof.

The drone operator expertly maneuvered the drone to a hover over Matt and Dave as he adjusted the altitude and let the gyroscopically stabilized camera focus in on the target. His eyes darted between the camera feed and the trigger toggle as he made a few final adjustments to his aim as the targeting app zeroed in on the operators. The trio began to chant, "Allah Akbar. Allah Akbar. Allah Akbar!" He slowly moved his finger to flip the toggle to release the drone's ordnance as the app's targeting system calculated for wind drift and applied the final corrections to the drone's drop position. Dave and Matt were completely exposed and the operator could see Matt gathering his supplies. The drone operator smiled. He knew that he would soon kill these infidels and show America that there is a high cost for occupying the caliphate's land. Staring at the iPad screen as he flipped the toggle plugged into the side port, he could see every detail to include the puff of smoke from Dave's rifle. He never finished the movement or had time to contemplate he had just watched his own death. The bullet Dave fired a hundredth of a second before he released the bomb ripped through the base of his skull killing him instantly. Equally unknown to Dave and Matt was just how close they were to being killed. As the lifeless operator fell to the ground, so did the iPad, which hit a sharp corner of concrete and shattered cutting the feed to the drone. Following up the first shot, Dave, known for his ability to rapidly fire accurate shots, quickly ripped off a second and third round killing the other two terrorists. Pausing for a moment, Dave sent one final round through the iPad effectively destroying it. "Glad the morons didn't spend the money for a protective case...you know, an Otter Box or some shit that they advertise you can run over, flush in the toilet, and shoot to the moon." Dave had no idea how right he was. Once the iPad was destroyed, the signal to the drone failed and it returned to its launch location as preprogrammed. Dave wasn't able to shoot it, but he could clearly see the drone disappear somewhere north of their position in a residential area. No doubt, wherever that drone landed, more would soon return. "Holy fuck! Are you happy Dave? Way too close,...but good shooting!

Now, if you don't mind, I am getting off the roof with or without you." Matt said. Dave looked up from his rifle. "I got them, we are good Matt. Take a deep..." Boom! Another deafening blast shook the roof, cutting Dave off in mid-sentence. The blast ended all further debate. Matt and Dave ran like hell. As they passed the edge of the roof, they could see what looked like two Marines lying lifeless around a blackened blast crater a short distance away from the gym. "Shit, they got more!" Dave and Matt didn't pause even for a second. Both men continued their sprint with their gear through the door leading off of the roof and back down in the building. Dave slammed the heavy metal door shut behind him and leaped down the steps. As Dave rounded the corner of the first flight of stairs another blast ripped through the roof of the building spraying them with concrete debris and throwing the operators against the wall. "Dave, you okay?" Matt called back. "Yeah, I have a mouthful of dirt, but otherwise intact." Dave answered. "Is it me or are the blasts getting a lot bigger?" Matt looked at the gaping hole in the roof and answered. "Definitely bigger. They are straight up bombing us now. Let's keep moving before the next one hits."

Meanwhile, back in the gym, the Gunny had pushed one fireteam back to the GSO warehouse to link up with the team still there holding the ground floor. "Pull back to the warehouse," Gunny ordered the Marines. Moving like they were trained, the young Marines began bounding back in pairs to the warehouse. "They are moving too damn slow," John thought. "Gunny, they need to run. Tell them to run. The hell with bounding." John yelled. It was too late. The blast sent everyone diving for cover. Another bomb was dropped nearly on top of the last two Marines killing them both instantly. This time though, no one heard a thing. The explosion seemed to come from nowhere. The terrorists had to be using bigger drones that were flying higher than any of them could see or hear. It was futile to look up anyhow because you were blinded by the sun overhead. John's mind kept racing to put it together. "Smart...put the sun to your back when you attack. They used a trick as old as warfare itself." James grasping the situation crawled over to John. "John, we can't move. If we come out from cover we will be hit. The bastards are sniping us with drones." James was right. The team was effectively pinned down. "Sir, we got to go out and get the Marines." Gunny asserted. "This isn't the movies Gunny. They are gone. We will come back for them, but if you go out there now, you are dead." James bluntly laid out the facts. "But sir," Gunny started to say. James cut him off. "No." John grabbed his head and closed his eyes to focus. "How can we beat these guys?" He knew they were dangerously exposed in the gym without support. It would only be a matter of time before the terrorists ran more vehicles modified into massive guided Tesla bombs into the building and leveled it. If they stayed put, they were dead. If they moved they were dead. There had to be a way out. "Gents, how many smoke grenades do you have?" John suddenly asked. The group had at least nine in various colors. "Okay, we have one shot at this. If we pop them all and time our movement, we have a shot of getting at least back to the GSO warehouse. Whoever has the best arm needs to wing that bitch as far as they can to make sure we still have smoke to cover our movement on the far end. You guys good with that?" The team considered their options and quietly nodded in concurrence. John still pondered for another minute, waiting for some sexy plan to be devised, but using smoke to obscure was about all they had. "Okay, get ready to move."

"Now!" John signaled the team to action. The doors around the gym flew open and a barrage of smoke grenades flew out in all directions. "Wait...hold on....just a few more seconds." John was watching for enough smoke to build to make an effective screen from any overhead drones. They had tossed the grenades out various sides to confuse anyone watching, which door they would exit. The green, white, red, and yellow smoke circled in the breeze as it began to rise and thicken into a psychedelic fog. Then the wind started to change and blow the smoke away from their position. "It's now or never," John thought. "Move!" The entire team sprang from the building. Not making the same mistake as the Marines, no one bounded. This was an all-out sprint and it was every man for himself. The men stumbled and tripped over debris obscured by the smoke, but pressed forward. There was no stopping. "Boom! Boom! Boom!" Blasts rang out from all around, but no one could see anything through the smoke. If there was any doubt before, there wasn't now that drones were overhead and ready to attack. If anyone went down, they were on their own. The team had already lost too many people and anyone caught in the open when the smoke wore off was dead. Erik and John emerged from the fading smoke leading the sprint. They were just short of the door. James and Mark, the team's resident gym rats were bulked up like Conan, but weren't known for their sprint speed and were lagging. "Move faster. Over here...follow my voice!" John yelled blindly into the thinning, but disorienting smoke. John lost sight of them, but continued looking into the fog yelling until a hand grabbed his shoulder and pulled him through the door. "We are up idiot. Get inside. Everyone made it inside, but you man." James' sarcastic voice was obvious.

6

Strategic Paralysis

"Where is the CO? Get the CO down to Ops right now. He has some high priority message from Washington." CO is the term for commanding officer. Colonel Wyatt was the 26th Marine Expeditionary Unit's (MEU) CO. The 26th MEU was a potent, self-contained military embarked on amphibious ships designed for rapid, global crisis response. It had everything needed to fight unsupported for at least 30 days to include jets, helicopters, tanks, artillery, and troops. "Attention on deck." Colonel Wyatt stepped into the combat information center or "CIC" for the situation brief. "Okay captain, whadda'we got? Give me the BLUFF (Bottom Line Up Front)." Turning to the CO, "Sir, at approximately 0810 this morning local, a complex attack was launched against the US Embassy in Tunisia, Tripoli. There appears to be significant damage and casualties that include US Ambassador Reese. The RSO has taken control of the operation on the objective and is requesting immediate military support. The attack is believed to still be on-going. The CIF team has been activated to respond, but they are in Sigonella (Sigonella is a U.S. Navy installation in Sicily, Italy). Their estimated response time on target is +6 hours. We are the closest unit available. Reapers are already being diverted from Niamey, Niger, but they are +45 minutes till TOT (time on target).” “What do we know about the enemy and friendly situation on the ground? Do we have direct comms established with anyone captain?” Colonel Wyatt asked. “Not much sir. It appears a blast occurred in front of the main embassy building and an SVIED took at the main gate. We know there is an MSD team on the ground now with green radio capability to talk VHF and UHF/Satcom and there is MSG Det. As of now, we don’t have direct comms established.” “Major Lee, how much time before we can launch the TRAP team?” Major Lee was the executive officer and had already prepped the contingency response plans for such a request. The “TRAP” team was a normal Marine rifle platoon that was specially trained and equipped to do “tactical recovery of aircraft and personnel.” Although not a true “special operations” unit, they definitely were given higher end training and were the only unit on stand-by around the clock that could be ready to respond within minutes. Major Lee stated, “Sir, the TRAP team has already been alerted and are mustered on the flight deck. They are ready and awaiting your orders to board Ospreys. The estimated flight time from launch is six and a half minutes to the objective. We are also ready to launch the Cobras and Harriers to provide air cover.” Cobras are what Marines refer to the AH-1Z “Cobra,” which is an absolutely lethal and combat proven attack helicopter, which in its earliest variant was used during Vietnam. The Harrier is the name for the Marine AV-8B, which is their primary ground attack aircraft and can take off and land vertically carrying a wide range of weapons and ordnance.

Even though the Harrier was being phased out and replaced with new F-35 Joint Strike Fighters, the older aircraft still had a few more years of service.

The CO took in the information. Looking to his 2, which is short for the S-2 or intelligence officer, he asked, "What's the intel picture? Who are we fighting?" The intel officer was in a rough spot. The best he could do is peddle some standard templated information about general threats known to be operating in Tunisia and the weather forecast for Tunis. As for what they were specifically dealing with, no one had answers. He had been on the high side, which is a term for the classified internet system used by our military and spy agencies to communicate, since the word of the attack broke, but no one knew who to blame and everyone was in a CYOA (cover your own ass) mode.

Langley was silent. The CIA was caught off guard again and didn't want to admit it. They would never be blamed openly, but this failure was squarely in their purview. No doubt, they would be pushing to blame the Department of State for the failures knowing full well, irrespective of State's security measures, State relies on intelligence from the CIA to drive what level of security is appropriate. Nonetheless, as soon as someone could safely attribute the attack, the CIA would be quick to establish that the group was their turf and they were the right organization to deal with the problem. Ironically, no one seemed to have connected the obvious dots between CIA failures and the growth and spread of radical Islam. In short, the question that Congress and the media desperately needed to ask was; if the CIA was doing so well, why in the hell were we losing the war against Islamic extremism? Using a simple process of logical deduction, there were only two possible answers: Either the Agency wanted to maintain an enemy of the US for nefarious purposes or they weren't up to the challenge. In spite of this, the CIA continued to maintain their coveted ownership of all things terrorism. After all, it made them relevant and influential, but more importantly, it got the organization billions of dollars.

For their part, the FBI had already ruled out the groups they were tracking. The FBI was primarily domestically focused, but because of the extensive overlap between domestic and international terrorist groups, it was nearly impossible for the FBI to be completely divorced of investigation beyond America's borders. In fact, conducting investigations overseas is why the FBI had a liaison officer in Tunis. Nonetheless, the FBI genuinely had no information to support a coordinated attack of any type was in the works against any diplomatic facility in Tunisia. However, in their case, denying any knowledge of the attack wasn't problematic. In fact, it was doubly beneficial for the FBI. For one, it was the CIA's job to track, collect on, and take out foreign terrorist organizations, which in this case, provided an easy out for the FBI. The FBI wasn't about to take the fall for this intelligence failure, which was how the Director had already decided to politically frame the situation. After all, intelligence collection wasn't their primary job. It was "investigation." As such, the CIA was automatically the patsy for this failure according the FBI. Second, there was a long running zero sum game between the FBI and CIA for turf, budget, and ultimately power. The FBI knew that any intelligence failure they could pin on other agencies would make them look better and drive Congress to allocate more money for the Bureau. As perverse as it may be from the viewpoint of the commercial business world, Congress actually rewards failure with more money, not less. One could argue this incentivized failure or at least the blame game rather than better cooperation between government organizations. The worse the attack was, the more money and power it meant for the FBI. From a Machiavellian perspective, like 9/11, this was going to be a great thing for the FBI. Winning even while America was losing, but no matter!

The DIA was enjoying the fact that most Americans didn't even know the agency existed. Under the circumstances, DIA was more than happy to maintain this anonymity even though it had a substantial budget and staff based literally around the world. Should it come out that they were also unaware of any threats, the canned response is that they collect and analyze intelligence on foreign militaries rather than terrorist organizations. Of course this was a half-truth, but it would stand up to Congressional scrutiny. Should they become highlighted, they would revert to the classic organizational fallback and claim they needed more resources. In any event, DIA hadn't even responded to mounting intelligence requests. Their attaché in country apparently had reported in, but was caught out in town when the attack began and wasn't able to get to the embassy. It didn't matter anyhow because he didn't have any information either.

Naturally, Colonel Wyatt wasn't thrilled, but had expected the confusion. Rather than waste time demanding something that didn't exist, mainly solid answers, he ordered that the MEU prepare for a rescue mission. His Marine training had groomed him for a bias towards action during the chaotic and fast pace of events that inevitably come with battle. This vacuum of information is referred to as the "fog of war" and American military officers are trained to act without perfect information. "Major, go ahead and get the TRAP Team loaded and have them wait on stand-by in the birds. Keep the flight deck clear so we can launch on my command." "Yes sir" was all the XO said as he got to work.

The CO needed a direct, secure line with Washington. No matter what his gut told him he needed to do, he couldn't launch the rescue without clearance from the White House. Sending combat troops into a sovereign nation was tantamount to a military invasion unless invited in by the host nation. Right now, Colonel Wyatt had permission from neither Washington nor Tunisia. "Sir, CENTCOM is ordering you to hold. They are trying to work with the host nation right now to use local assets." One of the operations officers relayed. "If the host nation was going to be helpful, they would have prevented the attack in the first place. We could be waiting all week for the Tunisians. Hell, their presidential palace is under attack and they can't even deal with that so what makes CENTCOM think they will be able to help the embassy?"

Turning back to his communications officers, the CO ordered, "Get me someone who can make a damn decision. I don't care; the NSC, White House, someone."

Exactly fourteen minutes later, the CO reemerged from his consultations with Washington. It was clear there was not going to be a decision anytime soon. The State Department was adamant that the Tunisians were the responsible party to provide help and that the military should not "invade." The only problem was no one in Tunisia was answering the phone because their entire capital city was under siege and the President was taken to a secret location for safety. It was apparent State had opted to blame Tunisia and was naïve in thinking they could evade the intelligence agencies hanging this on their heads. Either way, State was in denial about the disastrous fate of the embassy if the US did not send help immediately. Colonel Wyatt's low opinion of the diplomatic corps wasn't changed by this equivocation. It was the natural programming of diplomats to be vague and non-committal. Not only that, but what was State going to do anyhow, send in more bureaucrats? However, he was furious with the CIA. The CIA had legitimate assets in country that could help, but was less than forthcoming when it came to offering their support. In fact, the only thing the CIA seemed convinced of was that they had a single insular focus on "their" best interests irrespective of America's national interests. Getting involved was organizationally too risky. Their calculus was cold, but clear. CIA's leadership, irrespective of the guys on the ground, wouldn't compromise their sources, assets, or careers for the lives of Americans. Tax dollars well spent.

As for the White House, they knew they had a political disaster on their hands and were already more focused on how to spin the narrative than saving the remaining people at the embassy. The White House needed a fall guy and everyone knew it. The President had been loudly touting his success decimating terrorist organizations almost from the day he took office and had enjoyed quite a bit of popular support from the sound bites. He had surrounded himself by war hardened generals and repeatedly told America that this time the gloves would come off and the military would be allowed to take the fight to the enemy. America would destroy the terrorists the President promised. The military of course loved this, but their strategic shortcomings both politically and militarily would cost them dearly. The problem for the military was now they had no excuse for losing. They could no longer pin their strategic failures on the civilian politicians. Their strategic inadequacies and incompetence were going to be laid bare before the public, which had long since grown wary of spending trillions of dollars on perpetual global war that the military proved just as incapable of winning as the CIA. To use a Vietnam analogy, the attack was shaping up to be the Tet Offensive of the war against radical Islam. America would lose faith in the President and the military and want to wash its hands of further conflict. Instead of wanting revenge, Americans wanted out. The nation was bankrupt and the Orwellian police state created since the attacks on 9/11 had become oppressive. People just wanted to go about their business without being tracked, searched, analyzed, and generally treated like a criminal by their own government. This awakening of the public terrified the very real military-industrial-complex that plowed literally millions of dollars into political campaigns for billions in profits. Never ending war was critical to their wealth. Further, the President knew that his political future was now in jeopardy. The attack was going to be used by his opposition to destroy him on his foreign policy and counter terrorism efforts. Worse, insiders knew that the current Administration's policies that turned a blind eye to Saudi Arabia's continued support for Salafists wouldn't be given the same media blackout that President Bush had been able to enforce after 9/11. This cozying up with Saudi Arabia was considered a necessary evil to contain Iran on Israel's behalf, but had been backfiring since the moment the evil deal was struck. Evil begets evil. Don't ever forget it. The US was again relearning this hard lesson and the savvy Washington insiders were trampling each other to distance themselves from the President's policy. The fear was real and was quickly growing into a full scale panic as word began to leak that the President had ordered the diplomatic mission in Tunisia to remain even after intelligence clearly showed terrorist activity in Tunis had risen to dangerous levels. It wasn't that anyone wanted to see Americans killed; it was just that the damage was already done and the President's staff knew all too well they needed to get ahead of the story. Damage control was the order of the day and everyone wanted to pass the buck to cover their asses politically. This was the sign of the Millennial Generation ethos, which had permeated American culture to the very top. In 21st Century America, no one was ever wrong and personal responsibility ceased to exist. It wasn't about right or wrong, but how one felt. Reality could be denied since the State insulated even the dimmest of morons from themselves. From the individual to organizational level, career advancement was based on how well you could blame one's own incompetence on others. To say this had a detrimental effect on the moral, efficiency, and effectiveness across the government would be a severe understatement. As it turned out, today was the day reality had its revenge. You can deny reality, but you won't be immune from the consequences.

Colonel Wyatt, with the dignified grace of an old British Officer, quietly took a seat without uttering a word to his subordinates. As a good man, his conscience weighed heavy. One may complain vociferously up the chain-of-command, but not down. He knew good Americans were going to die and that his hands were tied preventing him from doing a damn thing about it.

7

Hold Out

The men were under siege in the GSO building and had settled in the best they could while awaiting help and trying to devise an escape plan. It was impossible to move outside under the bombardment. The sunny day had become gray from the black smoke, dust, and fires. The pungent smell of burning fuel and plastic permeated the compound. The immaculate court yard and embassy grounds were now indistinguishable from the hundreds of other cities across the Middle East and North Africa blown to pieces by recent wars. The courtyard was pockmarked with blast craters and strewn with debris. The iconic Date Palms ,whose fronds were just hours earlier waving in the summer breeze and heavily burdened with ripening fruit, now lay splintered across flower beds raggedly tilled by high explosives. Where high perimeter walls once stood, piles of concrete rubble marked their foundations. Even the animals could smell the death. Scavengers converged on the embassy. Packs of mangy feral dogs fought with giant buzzards over the remains of the dozens of dead that were now bloating in the summer heat.

The best the men could surmise was that the terrorists were use bigger drones flying at higher altitude because they were dropping much larger bombs and they could neither see nor hear the UAVs. Either way, the sorties seemed endless. The UAVs would hover over the building, drop their ordnance, and then fly off to rearm. Unlike manned aircraft that took hours or days to be ready for another mission, the drones had almost no down time between sorties. The terrorists had established a clandestine network of logistical bases throughout the city to support the UAVs. All the UAV had to do was fly back to one of the locations and a team would swap out the battery and clip on a new explosive. Then it was back in action for another attack. It was that simple. Altogether, the UAV "pit crew" took no more than a minute in total to turn a drone around and the flight time to the embassy was less than a few minutes from any one of dozens of landing zones. This rapid refitting of the drones allowed the terrorists to continue to relentlessly blast away at both the GSO compound and the embassy. Fortunately, the embassy was built to far higher security standards and was holding up under the barrage, but the GSO building was quickly turning into rubble. It was clear they wouldn't last much longer inside the building and needed a plan.

As the bombardment continued, McEwan had managed to consolidate everyone on the first floor of the GSO compound aside from the lookouts. It was risky, but they still had to take turns watching for a manned attack from two small windows framed into the concrete block walls that gave a good view of the two primary avenues of approach. One overlooked the now obliterated gym and main entry control point. The other looked back across the courtyard towards the embassy and side gate. The men fortified the rooms with whatever they could find that would slow or stop a bullet or blast. Book shelves, boxes of copying paper, desks, you name it. They piled it on to create a buffer. At least the rooms with windows had a breeze coming through even if they were more exposed to attack. Everyone else was suffering since the building no longer had electricity. As the building baked in the North African sun, the heat and stench began to rise inside the building no longer cooled by industrial air conditioners or serviced by functioning toilets. The temperature soon was stifling and the air thick. All of this made the already small room the survivors had hard pointed in just that much hotter, stuffier, and darker. However, it was the safest place they could find and staying alive mattered more than comfort. It was an internal room with no windows buffered on all sides by hallways or rooms. This gave enough protection to stop most of the frag from bombs detonating alongside the building. However, there was simply no air flow and the addition of the local staff and their requisite body odor gave the room the unsavory aroma of stale pizza. "Well, at least we will get some light and fresh in air in here soon if this keeps up." James said staring at the ceiling as it shook from the latest wave of blasts threatening at any time to blow through into the room. Dust and bits of concrete fell on the men's heads with each successive blast. Erik, also staring at the fractured ceiling said, "Yeah, do you remember the old movie Das Boot about the World War

Two German submarine crew? I suddenly know how they felt getting depth charged."

"Do you think anyone from the Annex is coming to help?" Matt asked. John turned to Matt and with zero inflection simply said, "No." "Well, that was pretty definitive." Dave injected. "What makes you so absolutely confident we are on our own?" Dave continued. John standing said, "I will give you a box of reasons. For starters, I talked to Grizz at the gym yesterday. The GRS guys were heading out of town last night for a few days so aren't anywhere close to Tunis. Second, the CIA isn't going to bail State's asses out again like it did in Benghazi. Remember, those guys weren't even supposed to respond. Thankfully, they disobeyed orders. However, the CIA has made it abundantly clear that shit like that won't be tolerated again. Finally, there is no way in hell they can get to us. Anything that moves is bombed. I am quite certain that they have their own problems right now to say the least." As if the others hadn't already figured as much out, James added, "Well, I have to admit, we are kind of running out of options. Seriously, I don't mean to be the Negative Nancy here, but we are really looking more and more fucked without any lube in sight. Good ideas are going at a premium right now."

John stared into the hallway as the building shuttered under another blast. It sounded like the top floor had just been pancaked by the latest barrage of bombs. "We got to get in touch with some heavier assets, John stated." "Matt, do you have the UHF antenna with you?" John asked. Matt scrounged around in his ruck for a moment pulling out seemingly endless amounts of gear before producing a beer can style antenna for satellite communications or Satcom. "Here." Matt said. Looking back at Matt, John remained in place and didn't take the antenna. He said, "No, you take this damn thing back and stick it out a window. You're the comm nerd. See if you can get a good ping on the green radio." The green radio was a term for the AN/PRC-152 Multiband Handheld Radio (Harris Falcon III). Relatively speaking, it is a portable and compact tactical radio that is capable of doing the job grunts used to have to carry multiple radios for and is manufactured by Harris Corporation. By switching to a UHF antenna and enabling a UHF frequency, the radio was capable of Satcom transmissions, which could reach anywhere in the world. As such, if they could get a strong enough signal from an overhead satellite, they could contact the MEU or other military units. "Dave, while Matt plays with the radio, take Mr. GSO and see if you can get any of the vehicles in the garage bays to start up. I am thinking if we can get even one to run, we can make a break for the embassy." "Only if I get to drive," Dave said. Dave was probably their wildest driver and that is exactly what they needed. "Fair enough. Deal." John said.

Matt dug out what looked like a map and a protractor. The device was similar to a slide rule and helped the operator locate satellites in range wherever you were in the world. If you used it properly, it would give you a rough azimuth and angle to point the antenna into the sky. Matt fumbled between trying to balance the antenna with one hand and aiming a compass with the other while trying to watch the signal strength on the AN PRC-152. This went on in vain for close to ten minutes. As expected, the "fuck this" moment arrived when Matt knocked the antenna out of the window. As it dangled by its coax cable, Matt threw the compass back into his ruck with a few choice explicatives and slowly slid his back down the wall until he assumed a seated position on the floor. John watching the entire ordeal just started to laugh, which was about to piss Matt off. However, John preempted Matt's outburst of frustration and clarified what was humorous. "Check out your signal," John said. Matt began to laugh too as he looked at the screen of the radio. "Ninety percent signal strength! I'll be damned. If I knew it was that easy, I would have kicked the thing out the window a long time ago." John jumped up and grabbed the radio. "Yep, it's good. Don't move." John said. Keying the radio, John began his traffic. "Any station, any station, this is US Embassy Tunisia. Any station, any station, this is US Embassy Tunisia." John waited only a moment. "US Embassy Tunisia, this is Baghdad TOC. Go ahead." Both Kabul and Baghdad had a good habit of widely monitoring Satcom frequencies for just this type of emergency.

"Baghdad TOC, this is Mongol One from the US Department of State. The embassy has been under a sustained complex attack and has sustained heavy damage. We have dozens of casualties and are currently holding positions in the GSO warehouse and the main embassy building. We are pinned down by enemy using unmanned ground and air vehicles. We need support immediately. Over." John released the mic hoping the transmission went through in its entirety as he gathered his thoughts for the next transmission. "Roger Mongol One. Can you authenticate the information? Over." Baghdad responded. "Baghdad TOC, no, I don't have a way to authenticate what I am telling you. Try turning on the news. We are pinned down in a building getting rapidly blasted to pieces. Confirm it with Washington. In the interim, can you start spinning up some assets to help?" John blurted out in frustration. "Mongol One, we copy and will relay to higher. Can you say again on the enemy situation? We copied you were pinned down by...uh, drones. Over." Baghdad replied with a bit of a sarcastic intonation. John tried to be more specific. "Baghdad TOC, that's affirmative. We are dealing with something totally new here. The enemy is using drones to attack the embassy with virtual impunity and has also launched manned attacks using drones for covering fires. We were able to stop the manned attack, but not the drones. They are bombing anything that moves. Over." "Copy that Mongol One. Have you seen the drones? Over." John could hear the skepticism in the watch officer's voice and was visible perturbed. He knew where this was going. They were in denial and wanted to blame something like traditional rocket and mortar fire for the attack. "Baghdad TOC, this is Mongol One. Yes, I can confirm that we found a remote and various small drones in a pack on a hostile KIA'd. We also saw a small UAV fly back into the city after an attack. They are hard to see though. They are small and flying high. Over." Baghdad's skepticism continued. "Mongol One, we copy you found drones in a pack and saw one fly over the city. Can you confirm that they were armed? Did you actually see any drone attack something? Over." Without hiding his anger, John rekeyed the mic. "Baghdad TOC, no, we cannot see the drones actually attacking because they are flying too high and backlit by the sun. However, we can hear them if they fly over low, which has corresponded to getting hit with bombs. We also identified and engaged hostiles remotely controlling a drone. Over." There was a long pause on Baghdad's end before they replied. "Mongol One, this is Baghdad TOC, we will relay your information and provide you updates as we receive them. Try to sit tight and hold out. Be advised that engaging personnel with a just a remote control isn't in accordance with ROE unless you know for sure they pose an imminent threat. We will continue to monitor this network and conduct radio checks at the top of each hour. Over." John closed his eyes and slowly exhaled as he calmed his temper and gathered his composure. "Roger that Baghdad TOC. We will try to consolidated all personnel at the embassy and await support. Out."

"They think we have lost our minds don't they? Matt asked. McEwan was still hot. He knew Baghdad thought they were panicked and seeing things. On top of that some POG (people other than grunts...a.k.a. admin clerks) in the command center had the nerve to second guess Dave's decision to engage a threat. Few things pissed John off more than someone thousands of miles away arm chair quarterbacking a fight. In McEwan's book, this was up there with the Prima Donna of a general, David Petraeus. Petraeus, or "betray us" as he was known by many troops, was infamous for berating his CIA staff if his strawberries weren't cut at the perfect angle. Far worse though was at the DoD where he developed a bad reputation for denying fires for troops in contact while sitting in Florida, which got many good men killed. John finally answered Matt. "Yeah, they think we are idiots. Their pea brains can't comprehend that something could be happening that we haven't previously dealt with. This is really bad because when help finally does arrive, they are going to dumb into this kill zone and get eaten up too. For now, we are on our own."

"John, we got two FAVs to turn over," Dave reported. "Awesome. Can we fit everyone in them? This building won't take much more." John said. "The one is a sub (Chevy Suburban) and the other an LC (Toyota Land Cruiser) so if we take out the seats, we can probably jam most in the sub if we pile in like a clown car. The rest will have to fit in the Toyota. We only will get one shot at this move." Dave said. Before John could answer, he heard gunfire from the room with the window facing the entry control point. Running into the room, John and Dave saw Erik firing into what looked like a D-9 bulldozer methodically making its way towards them through the rubble. "Oh shit, don't tell me that is another unmanned VBIED." Dave said. Erik just nodded as he continued to pour as much fire as he could on the vehicle with Dave and John joining the engagement. "We need something with more punch. These rifles can't penetrate the armor." Erik yelled as he changed a magazine. The 5.56mm rounds of their Mk 18s and M-4s simply didn't have enough power to penetrate the steel plating rigged onto the homemade, but effective armored vehicle. As they continued to try and find a weak spot to disable the vehicle, it slowly plowed its way closer and closer. The obstacles formed by the rubble they had assumed would insulate them from the massive bombs carried by vehicles suddenly was about to be breached. "Get Mark up here with the 203," John yelled. "If that dozer makes it to us, not only will it take down the whole building, but can punch into the embassy and plow a path for other vehicles." Mark Neise came running in and took up an offset firing position beside Erik with his M203 as John and Dave rolled out of position.

The M203 carried by Mark was a 40mm grenade launcher and packed the most punch of any weapon MSD was able to deploy with to Tunis. In fact, this weapon system wasn't even officially approved by State and the Department had never approved or stocked lethal ammunition for it. Nonetheless, operators still maintained a few M203 systems with ammunition from battlefield acquisitions. This lack of weaponry was a serious sore spot because the operators were trained on and required a host of weapons much more powerful than the M203 such as dedicated antitank weapons and machine guns. The inability to bring additional weapons on missions left the units almost always operationally outgunned and was a self-inflicted wound by the Department's management. The Department hated the idea of guns in general and resisted the deployment or carry of anything more potent than the basic M4, which itself took over a decade for management to embrace.

Even though the Department needed a Tier 1 paramilitary unit to mitigate high threat/high risk situations, they clearly didn't want it and lived in a constant state of denial. It is kind of like needing a good plumber, but hating the idea of anyone working on your pipes and forbidding them from carrying a wrench or plunger. The logic just didn't compute, but the de facto reality was that a team was limited to a logistical footprint no bigger than what an individual could carry. To even a typical civilian, which can't fly anywhere without three pieces of luggage, this was a mind blowingly unrealistic constraint. In practice, the Department's logistical support system made it simply impossible to deploy with anything bigger than a compact rifle. If it wasn't in country already, it wasn't getting there via State. To summarize how this operational reality came to be, the State Department's leadership had been indefinitely living in utter denial about the current and emerging threats faced by operators. Specifically, Washington couldn't shift gears between the security policies necessary to effectively deal with the different threats posed by insurgent and conventional state-level capabilities. Apparently, Washington had typecast all threats at the insurgent level and never adapted past the threats in Afghanistan and Iraq. However, in countries like Libya and Syria, full scale conventional weapons of war were being used such as tanks and aircraft, which required heavy weapons to defend against and knock out. If the MSD unit had even one AT-4 (84mm disposable antitank rocket), the dozer could have been quickly dealt with, but under the circumstances, the M203 was their best bet and it was still a poor choice from a weaponeering perspective.

The distinctive, dull "bloop" sound of the M203 firing caused the others to cease their vain attempts to knock out the dozer with their rifles. Everyone watched as a golf ball sized dot arched toward the dozer, landed a few meters to its side, and detonated with a loud explosion showering the side of the dozer with shrapnel. "Yeah!" the guys roared when the grenade detonated, but then they got quiet. Despite its proximity, the dozer didn't seem phased and continued to slowly plow forward. They were dead if the dozer made it another 50 yards. "Damnit, the angles are bad. It has too much steel up front. We need to get a shot on the side or from behind to penetrate the armor. Even with a direct hit, it still may not knock it out." Dave said. Mark grabbed the bandoleer of 40mm grenades and slung it over his shoulder as he slid in another golden egg of death. The standard 40mm high explosive round got the golden egg moniker because the explosive warhead is literally a golden color. "I am headed out," Mark said. John tried to stop him. "Mark, that's suicide. Hammer it from here and drop a round on its hood. You go out and you are dead. Worst case, we pull back." Mark just shoved John aside as he headed to the door. "Get the fuck out of my way!" Mark was naturally a big guy with an intimidating build he worked on daily at the gym. The team had more than once speculated that Mark's size acted as a crutch during his early development and he must have failed to learn normal social skills like how to not be an asshole to everyone around you. Today the "I don't play well with others" attitude was on full display and no one was going to deter Mark.

Moving out of the door, Mark ran straight to what used to be another building the local employees used. Setting up in the rubble of the building, he actually had decent overhead cover. If a drone operator hadn't spotted him making his move, he was in the clear and lined up for a nearly perpendicular shot on the dozer. Taking aim from a knee, Mark fired the M203 and drove a high explosive round directly into the engine block of the dozer. The blast from the small shaped charge cut through the steel plate and into the engine compartment. The dozer belched blue smoke as the diesel engine blew out oil and fuel. Red hydraulic fluid also sprayed out in an arc from tubing severed by shrapnel spattering the dozer as if its artery had been severed. As it lost hydraulic pressure, the plow blade dropped and dug into the earth slowing the dozer. "Did it stop it?" Matt asked. "I think that did it. No, wait. Shiiiiiit." John stated. In spite of the damage, the beast continued to chug forward like a wounded buffalo. Just as the men were reevaluating what now seemed to be a premature sigh of relief, a loud mechanical clanking could be heard and the dozer stuttered to a stop. Everyone, including the huge dozer, sat motionless in an eerie silence. Then, by a remote command, the dozer erupted into a giant fireball. The blast sent a tremendous shockwave tearing across the compound, which scoured the ground and obliterated any last standing structures in its immediate proximity. Again, the operators were slammed backwards by a blast, but otherwise uninjured

81

because they had both standoff and good cover. Had anyone been in the open, they would have been blown apart. As the dust settled, Mark stood up and began shouting in victory at the smoldering crater where the dozer sat just moments before. "What the hell is he doing?" John asked. "Mark, get some damn cover you idiot." Dave shouted out the window. Mark looked up and began yelling as he fired his rifle into the air. "Fuck all of you. Bring it! Bring it!" A split second later they brought it as Mark doubled over to his knees as he cried out in pain. Neise was shot square in the balls by an enemy sniper. The gunshot was audible, but no one could identify the location of the shooter. This was real life, not Hollywood. You can't get away with dumb shit and Mark just paid a hefty price for hubris. The shot shattered Mark's pelvis and left him on the ground unable to move or even roll over. "Stay low man, we are coming to get you." Matt shouted. Mark was now lying in a pool of his own blood and bleeding heavily from the groin and pelvis. It was a horrible wound. John quickly grabbed his med bag and one of the few remaining smoke grenades in preparation to go recover the guy that just told him to get the fuck out of his way. Mark continued to cuss as he writhed in pain bleeding out. "Stay still man and get behind some cover," John yelled to him as he prepped the grenade and readied to make a final dash to his position. "What the fuck do you kn..." was all Mark said. He never finished. Another round fired from the sniper hit Mark directly in the face killing him instantly. John knew it was over for Mark and pulled back behind cover as he made his way back into the GSO building. Even though Mark was a complete dick to just about everyone, it was devastating to watch his teammate cut down in the street. Mark had saved everyone in the building, but paid the ultimate price for arrogance.

8

Fall Back

It was bad enough Mark had been cut down by a sniper, but he had the M203 with him. There was no way anyone could get back out to get Neise's body or the weapon. If the terrorists launched another dozer into the compound, there was no way they were going to be able to stop it. "Just pray they don't realize we don't have any other heavy weapons. I don't think we can hold out till dark. We need to go now. Let's load up." John told Dave. Dave yelled to the group, "Let's go. Remove the bench seats in the back and pile into the vehicles. Take your kit because we aren't coming back." While they were loading, Gunny called back to the Marines holding the embassy and had them prep the sally port door. The plan was simple enough. Everyone in the GSO warehouse would squeeze into two vehicles and then they would drive like mad across the open courtyard to the side utility entrance. As they approached, one of the Marines would open the armored bay door to the internal loading dock and they would drive in. As they entered, the Marine would shut the door and they would then link back up.

"Okay, now!" Dave yelled. One of the Marines lifted the garage door and ran back to the vehicle and jumped in landing on top of half a dozen other guys jammed into the back of the vehicle. As soon as the Marine was in he yelled "up" and Dave floored it. The FAV was a bit sluggish due to the weight of the armor, but nonetheless, still had the horsepower to launch them out of the bay onto the street. Dave had no problem burying the accelerator, which threw everyone into a dogpile in the back against the rear armored door. McEwan followed close behind in the second vehicle as they fish tailed around the corner. They hoped to be safely inside the embassy in less than thirty seconds based on their planned route. "Raise it," Dave told Gunny as they wound through a maze of debris and raced towards the embassy. "It won't go up Gunny. It's jammed." The Marine on the radio called back. The first two explosions erupted to the left of the vehicles. "They are on to us Dave, move it." John radioed. "What the hell do you want me to do, the door won't open." Dave radioed back. "Drive like hell and buy us some time. I don't think they can target us well when moving." John radioed back. Hearing this, Dave smiled taking a somewhat macabre pleasure in outdriving the drones. Dave reached over and turned off the traction control and placed the vehicle in a sport shifting mode. For the next five minutes, which seemed like an eternity, Dave and John raced around the compound staying one step ahead of the armada of drones dropping ordnance. The FAVs would sprint forward, lock brakes, and then reverse out the direction they came. At one point, the vehicles did donuts in the gravel kicking up enough dust to completely obscure both vehicles until they blasted out of the cloud toward the parking lot weaving in and out around the rows of parked cars blowing apart to their left and right as they continued driving. "I am going to be sick," one of the locals yelled before puking all over the backs of two other passengers. The donuts, turns, braking, and weaving were turning everyone's stomachs. Even the hardiest of stomachs was beginning to wonder if it would be better to take their chances with the drones as the stench filled the FAV. Nonetheless, as long as they kept on the gas, the swarm of drones couldn't quite hit them even though they had gotten pants soiling close. That said, both vehicles were peppered with shrapnel, John's windshield was shattered, and Dave was running on two flat tires. Apparently, no one ever checked to see if the mechanics swapped the run flats when the embassy had new tires put on the FAVs. It was now quite obvious that this small oversight was a bigger deal than anyone had foreseen as Dave's tires tore away in chunks leaving him limping forward dangerously slow on two metal rims. "McEwan, can you see what the hell is wrong with my vic?" Dave radioed. "Yeah, you don't have any tires left." John confirmed. "Oh, I thought it was something serious," Dave replied as he turned on the four wheel drive and locked the differentials to put power to all four tires. "I may not be able to turn as well, but I can keep it moving just like when we escaped from that bar down in Mexico on four flats." Dave continued. As is the norm, Dave spoke too soon. A wall of explosions bracketed the vehicles before six more sequential blasts walked right over the vehicles leaving Dave's vehicle unable to drive. One of the mortar rounds detonated right off of the front bumper and effectively took out his front axle and engine. Dave's windshield was so shattered it was almost impossible to see out of it and his side mirrors were completely blasted away. Dave tried to keep the vehicle moving forward, but it was down hard. "Vehicle down! Vehicle down!" Dave radioed as the vehicle plowed to a stop. The drone operators had figured out what the Americans were doing and how to compensate for their inability to target a moving object. Their answer was simple and effective. Mass and line up the drones over every single route and have them all drop their ordnance as the vehicles entered the kill zone. Most of the drones would miss, but the massed fires would almost guarantee a hit. "Stand by for push out." McEwan called giving Dave and his passengers barely a second to brace before John rammed their bumper and began pushing Dave's vehicle forward just as they had trained on tracks in West Virginia a hundred times. John wasn't being nice and gentle like in training though. He only slowed enough to prevent any serious damage to his vehicle. John had to keep a steady speed to keep good contact with Dave's bumper and taking any turns was less than ideal. They were now easy targets. However the blasts had noticeably ceased. "I think we have a window here Dave. They dropped everything on us and need to go rearm. Best I can guess is we have about a minute, maybe two, to get that damn door up and get into the chancery or we need to find another plan. Let's make a run directly for the sally port." John radioed.

"How much longer on the door?" John frantically asked over the radio. "I think we got it sir. Yes, it is going up. Get in here now." The Marine radioed as they finally dislodged the debris wedged into the door's track. John kicked his vehicle into four wheel drive and pushed Dave as fast as the FAV would handle as they made a second run on the sally port. Blasts were now beginning to bracketing their approach again as if they were making the final run on the Death Star in Star Wars. Dave radioed, "I can't see shit so call out the turn, but whatever you do, don't slow down." "Roger, standby." John called back. "Okay, on my mark turn hard left. 3-2-1-Now!" Both FAVs made the final turn onto the ramp sloping down into the embassy as sparks flew from the metal wheels and axles grinding into the concrete. They made no attempt to slow down. In fact, they sped up. "Shut the bay door." Gunny called to the Marines. The corporal hit the plunger to begin dropping the door. A second later, both vehicles came sliding into the bay with their e-brakes locked, wheels throwing sparks, and tires billowing smoke followed by another series of blasts just outside of the bay doors. "Oh damn sir." Gunny said as he looked back out towards the ramp. One of the drones had dropped to a low hover and was trying to maneuver under the door before it closed. "Shoot it!" Dave yelled as a dozen people piled out of his FAV half covered in vomit. John immediately saw the threat. "Don't let that damn thing in here. If it sets off a charge in here, it will open a major breach right into embassy." Immediately small arms fire erupted, but not a damn person seemed to hit a thing as the drone darted back and forth and then flew straight toward the bay. James dropped his Mk18 and swung his breaching shotgun into position. Leveling it on the drone as it raced directly toward him he cut loose blowing a rotor off of the UAV at nearly point blank range, which sent it spiraling into the wall. "Die mother fucker!" James yelled as he pumped the shotgun. James fired again and again literally blowing it right back outside under the bay door as it finally closed. John leaned back against the side of the FAV like he was conducting a trust fall. "That was too close, but it's about time we found a use for that breaching shotgun. Good shooting." John said as everyone took a deep breath.

9

Help is Coming

The embassy was a defacto fortress. It had been repeatedly hit with bombs all day, just like the GSO warehouse, but the thick fiber reinforced cement walls tied together with a lattice work of steel rebar was proving tougher than the ordnance the small drones were capable of carrying. Bombs would hit the thick concrete roof and the explosion would blow out and away. The explosions failed to blast through the building and left little more than divots and black scorch marks. They had, however, destroyed all of the communication antennas mounted on the roof. Coupled with telecommunications being down in Tunis, this reduced the embassy's communications to intermittent cellular calls and once they had set up another antenna. The city's communications were cut after preliminary drone strikes took out key communication nodes around the city. As such, Satcom was now the primary means of communication. Still though, this was enough for the embassy to reestablish contact with Washington and start passing vital information and planning a rescue mission.

Safely barricaded inside, the rest of the embassy staff had been productive. Wes had most of the casualties stabilized. The Marines had cleared the building and Jennifer had assembled and accounted for all of the embassy personnel. Anyone that needed a job was helping shred, burn, smash, or otherwise destroy all of the classified material they could. The RSO, Bill Burnett hand already given the destruct order for all classified. It was obvious to him at this point things weren't going to get better. While that was going on, Bill had stabilized enough that he was able to contact Washington again and give the Command Center a full update, which was now buzzing with activity and senior leadership. The Air Force had also retasked a reconnaissance drone, which was now flying circles high overhead monitoring the on-going attack to provide the military with situational awareness. The drone feed had become Washington's primary source of information in real time. Based on the current information, Washington had apparently settled on a rescue plan. A Marine TRAP team from the 26th MEU floating less than 30 miles off the coast was going to be sent in after nightfall to set up security and extract embassy personnel. They would use the interior court yard as a landing zone known as an "LZ" even though both the RSO and MSD team recommended against it. The embassy staff had already witnessed the drone capabilities first hand and had little doubt that the enemy would be prepared to continue the attack into the night. Irrespective of the warnings from the folks on the ground, the military felt confident they could clear the airspace of drones and hold security long enough to extract the embassy personnel. Either way, no later than 2100, everyone needed to be staged and ready to go.

Back in the States, the DS Command Center had now transferred control of the operation to the White House where they had stood up their full situation room and ops center. Everyone of rank to include the President was now huddled around an array of screens watching live CCTV footage from the embassy being piped in from the DS Command Center and the drone feed. They were helpless to do anything at the moment beyond watch small drones whir across the city and embassy grounds bombing targets with impunity. It was clear where at least some of these drones were taking off and landing and the Air Force was already working as quickly as possible to use all resources and intelligence to establish a target list to take them out. However, politics as usual was slowing the response. No one would clear strikes until precautions were taken to minimize collateral damage, which was proving nearly impossible in the densely populated downtown areas of the city. The terrorists were clever and had picked their launch and recovery sites very carefully. The all flew from heavily populated areas and often took advantage of Mosques, schools, and hospitals. Further, the precise site locations were ambiguous. No one had wargamed the fact that small drones could fly in through something as small as an open window. Once inside a structure, the drones could be flown through a building to a completely different area. For example, one site appeared to be in a major hotel's parking garage allowing the drones to fly into and through the open air structure.

The exact location to target in structures such as the garage was impossible to identify from the air due to the overhead cover. To take the locations out, the US would have to drop bigger bombs. The problem of course was this made it nearly impossible to take out the positions without killing hundreds of women and children, which would become a propaganda victory for the terrorist's recruitment efforts against the US.

10

Rescue in the Night

As darkness blanketed Tunis, the city assumed an eerie state of calm. The normal bustle of a city was frozen in fear. No one knew if the attack was over. Inside the embassy, the defenders were low on ammo, bloodied, and exhausted. The on-going battle had reduced the vibrant and picturesque embassy compound to something akin to a blown out city block emblematic of the conflicts plaguing the Middle East. The high concrete masonry walls of the embassy were now piles of rubble impaled by twisted pieces of rebar. The structures still standing were pockmarked with holes and stained black from soot. Carcasses of dozens of vehicles in the motor pool continued to burn as well as numerous other fires around the compound, which provided a hellish glow in the darkness.

Dave continued to scan his sectors for movement. He had been running on adrenaline all day, but fatigue was becoming overwhelming. Fortunately, James had commandeered a Nespresso machine from the Caleb Stroh's office, which was sitting neatly by itself under a picture of Stroh trying to look hard with a hipster beard and body armor. From the look of the photo, James guessed it was from when Stroh was in Syria, which was information Caleb always was quick to volunteer. However, Stroh seemed to forget the part that he was a total train wreck and nearly derailed his stabilization mission. James never got the full story, but the rumor network said Stroh was so insecure he overly compensated by not allowing any necessary support into the country. His obsession with control ended up alienating everyone so badly the Bureau of Near Eastern Affairs had to recall him after he had a nervous meltdown. Either way, no one cared about Caleb. What mattered right now was that somehow those little shots of heaven made even this day, which had become a hot mess on top of a dumpster fire, seem tolerable. Still though, in spite of the persistent dangers and a steady stream of Nespressos, Dave's eyes were getting heavy and he felt himself nodding off. Dave continued to reassure himself as time wore on. "Just keep focused. The Marines will be here any minute." Dave completed another methodical sweep of the compound. Quiet. The grounds had been deathly still for over an hour. Just the same green and black images that he had been staring at through his PVS-31 night vision goggles all night. "Ah, why do I have to have discipline? Can't I just say screw it and go take a nap?" Dave quipped to John. John was having his own thoughts that mirrored Dave's, but rather than reinforcing the misery, he just gave an acknowledging "hmmm." As Dave began to ramble on about his eyes playing tricks, John caught what he thought was a flash as he stared through his goggles to the north of the compound. At first he thought it was an anomaly, but very quickly he could see numerous faint lights blinking in the distance. "Dave...Dave, look to the north. What do you see?" John whispered. "Hang on." Dave noisily readjusted on his Islamic prayer rug that doubled as his shooting mat. "Okay, what am I looking for McEwan? I see the city and a bunch of lights." "No smartass; look out at the skyline." John snapped back. Dave refocused on the horizon and quickly saw what John was referencing. "I see what looks like maybe six blinking lights getting closer John." "Yeah, that's what I thought. I think our MV-22s are finally inbound."

"Grab the green radio and see if you can raise them on the guard network Dave." Neither one of the operators knew for sure what radio net the inbound aircraft would be communicating on, but they knew that the emergency frequency, or guard net, for military aircraft was 243.0 MHz on their VHF/UHF radios. "Any station, any station, this is Mongol One over." The 'shhhhh' of static answered their call. "Any station, any station, this is Mongol One over." Then in crystal clear sound, the radio came to life. "Mongol One, this is Thunder. We are inbound to your location. Two minutes to TOT. Cobras are in escort and will clear the LZ. Over." John remembered from his days as a Marine down at Camp Lejeune that "Thunder" was the call sign used by VMM-263, an MV-22B squadron out of Marine Corps Air Station (MCAS), New River, North Carolina. A moment later, another air unit checked in. "Mongol One, this is Vegas. We will be overhead in less than one mic. Over." Vegas was an AV-8B squadron from VMA-223 from MCAS Cherry Point, North Carolina. The AV-8Bs are Harrier attack jets and carry a powerful payload of weapons for close air support. They were a potent addition to the battle, even if now an obsolete airframe, which combined with the Cobra attack helicopters, made for a fearsome team. The radio came to life once more. "If you have your strobes, now would be a good time to turn them on." Thunder instructed. John shook his head. Damn it. "Negative Thunder. Enemy has IR capability and will target the strobes. Come in dark and look for your LZ 50 yards to the North of the building marked with 4 chemlights. Over." Before dark, James and Matt and snuck out on the roof and pitched IR chemlights in each corner. Anyone coming to the rescue would need some help picking out the exact building they were in and the chemlights would work nicely. From the air, a distinctive square would be outlined that at least theoretically would be unique and visible to the aircraft above during the night. They figured that the bad guys already knew where they were and couldn't blast through the roof so it didn't change the game for the worse. Using the chemlights as markers also didn't telescope the landing to the enemy since they had been on the roof for hours now and kept the operators from having to expose themselves again on the roof to set out strobes.

Moments after the last radio traffic, the thunderous roar of jet engines overhead could be heard. Twenty more seconds ticked by and the whup, whup, whup of the Cobra attack helicopters joined the cacophony in the pitch black sky. The reassuring sound of the rescue team finally coming set loose wild cheering amongst the staff. The embassy personnel were at their wits' end and had been creeping closer and closer to the door leading outside from Post One all evening. They were like little kids waiting up in anticipation of Santa Clause, but of course scared to death. Upon hearing the aircraft, about a half dozen staffers jumped up in celebration and ran outside waving their arms. "Oh boy. Get back in here and shut the door." James yelled. The MSD members scrambled after them and vainly tried to round them up and get them back inside. As they shouted for the staffers to get back inside, the ominous buzzing of the drones returned. "Not good. Drones are back. We have to get inside right now." Matt yelled. The terrorists had been waiting for this moment to launch another attack. "Matt, Dave, James, do what you can to cover those idiots." John commanded. Hugging the side of the chancery and being careful to remain under the protective awning, the operators inched out using their thermal/IR fusion optics to try and locate the drones in the night sky. "Where are they at? Anyone see anything?" The operators called out as they pied off each section of the sky as if they were clearing a room vertically. "Got one." Dave yelled. "Where?" The rest of the team immediately asked. Dave pressed the pressure switch on his PEQ15 mounted to the end of his Mk18 and sent a laser only visible in the IR spectrum into the sky. "I'll paint it with the laser." Dave called out as the reflection of the laser off of the drone made the target immediately apparent. The entire team then keyed their lasers and began firing into the sky as the remaining idiots made their way inside. The entire episode had to look like a mini antiaircraft battery opening up as lasers and red tracers crisscrossed the sky. The Marine pilots seeing this as they approached weren't thrilled and began calling "cease fire, cease fire" to no avail. Perhaps counterintuitively, the drones turned out to be easier to see at night using the optics. Further, by using the lasers boresighted to the weapons, they were also easier to engage. Within seconds of the volley fire, sparks could be seen in the air as their .77 grain, 5.56 millimeter bullets ripped into the small drone hovering less than 100 feet above. "Yeah, down one drone!" Dave yelled as the drone spun out of control and crashed into a pile of debris on the far side of the compound. In response, the area around the embassy suddenly began erupting in explosions as the rest of the drones dropped their bombs. Once they discharged their ordnance, the drones immediately increased their altitude beyond effective range of the optics. "Boom, boom, boom!" The bombs began to detonate all around. "Get back inside. Everyone. Now!" John yelled. He didn't have to waste his breath though. The second the explosions started, everyone piled back inside in a human stampede. "Did we get everyone? Anyone injured?" John asked. "We're up. All good." Came the reply. Matt and the operators then walked to the corner of the room to chat. "Did you guys see how the other drones responded when we opened fire?" Matt asked. "Yeah, they all pulled up beyond any accurate fire we could throw at them. I think they know our capabilities." James answered. "I agree.

They won't give us another chance like that." John said. "Yeah, but what I am most concerned about right now is I don't think the Marines are going to be able to land." Dave added. John looked grim. "Your right Dave. No way they are getting in. We need to call them off right now."

The operators didn't know the half of it. While they regrouped inside, the high dollar, advanced Marine aircraft were proving completely useless against the small, cheap drones. RADAR wasn't effective in detecting the very small and low flying drones. If the radar could even pick out the drones from the clutter of the city, the return was smaller than a bird. Adding to the problem, the jet pilots were moving too fast to even see the small drones far below their safe flight altitude, which blended in with the city. The drones were also blacked out and not emitting any illumination...even in the IR spectrum. Even for the helicopters, which could fly slow or hover, the drones were able to fly in areas the helicopters could not. The drones darted behind buildings, down alleys, and under wires. In fact, it appeared that some drones may have simply landed and remained still proving to be impossible to pick out from the clutter of the city. The bigger helicopters were just not small and agile enough to be able to get a fix on a drone. Further, even if a Cobra could target a drone, it was a weaponeering mismatch to try and use guns, rockets, or missiles. Assuming they could even be targeted and hit, the attack angles and blast radius of air launched munitions guaranteed significant collateral damage on the ground. When the drones were hovering right over friendlies, it was impossible to attack them without killing the people on the ground under them. Collectively, these factors had actually given the incoming Marines a false sense of security. The lead attack aircraft had been overhead for two minutes and had failed to detect or identify a single drone and falsely concluded the LZ was clear.

"Thunder, Thunder, this is Mongol One. Abort, abort. I say again, abort primary LZ." John called out over the guard net. "Thunder this is Vegas, disregard Mongol One. We have eyes on the LZ. You are clear to proceed." John looked at the rest of the team in disbelief. "They are flying into an ambush and we all know it." Dave said. "Too late." Matt reported as the deafening roar of the MV-22 rotors drowned out the rest of whatever he said. The operators could see the first MV-22 assuming a hover over the courtyard through the ballistic windows of Post One.

The drone pilots already had their orders and had been waiting for this moment. They knew that the Americans would attempt a rescue sometime during the night and had pulled back their drones to refit and rearm them in preparation for their attack. Their orders were to wait until the helicopter landed and then attack. The noise of the aircraft overhead made it readily apparent the rescue attempt was underway, but without drones up, it was difficult to know the precise moment to strike. As such, their plan was to employ sniper teams around the embassy to act as forward observers. The teams were not the typical amateurs generally encountered in Iraq and Afghanistan. These shooters were legitimate, trained snipers. The teams had state of the art, Artic International, rifles chambered in .338 Lapua, which gave them an effective range of over 1500 meters. Having the ability to standoff from such distances gave the snipers a huge advantage. None of the weapons the MSD team was carrying could come anywhere close to reaching out that far. The snipers also had the same night vision capability American Special Forces used. This shouldn't come as a surprise because the US had given the technology to them when they were trained by US Special Forces and CIA Ground Branch officers for combat in Syria. The insanity of US policy, which had committed the US to arming and training known radical Islamists fighting in Syria, had now born its poisonous fruit and come full circle. Now well armed, trained, and equipped, the snipers were a lethal threat to the Americans. As the snipers observed the embassy from their hides, they could see the Marine aircraft approaching for landing. Calmly they waited for their moment to attack as they continued to passively observe the aircraft through their night optics. No more than a minute passed before the MV-22s flew right into the ambush. As the MV-22s transitioned into a hover over the LZ, the observers picked up their encrypted, two-way radios and signaled to launch the attack.

The LZ was too small to simultaneously land the MV-22s. As such, the first tilt-rotor aircraft came to a hover and slowly descended onto the courtyard blowing up a massive dust storm that browned out the area and turned visibility to zero in the dark night. The powerful rotor wash was magnified by the high walls of the compound and created what felt like a tornado. No doubt, it had to have made the landing very challenging. In fact, a similar situation was said to have caused a heavily modified Blackhawk helicopter to crash during the insertion of US Navy SEALS during the Osama bin Laden raid in Pakistan. If it wasn't for the dust though, the pilots may have had a chance to realize they were under attack and pull off. Hidden in the dust cloud, the first drones came in too close for the attack and were slammed to the ground by the rotor wash. Breaking into pieces, the drones were then blown across the LZ into a fountain. The drone operators were caught off guard by how much more powerful the rotor wash from the MV-22 was than a typical helicopter and were blinded by the dust when it enveloped their drones. Unbeknownst to the MV-22 crew or the aircraft overhead, they had involuntarily taken out the first wave of the attack. Unfortunately, they didn't realize it and take evasive actions. Instead, the crew chief dropped the ramp as a squad of Marines prepared to exit the aircraft and set up a security perimeter as they were trained.

"Can anyone see a thing?" John asked. "I can't see dick...and definitely not your little dick James." Dave answered. James turned to Dave and left it at, "your momma didn't seem to mind it." As the team peered through the thick ballistic glass into the dark night, the dust cloud suddenly turned into a bright glowing fireball. It had the appearance of how a cloud is illuminated from within by a lightning strike. Then the shock wave from the explosion hit the window sending everyone diving for cover as pieces of giant rotor blades slammed into the front of the chancery. "Son-of-a-bitch! They hit the Osprey." John yelled. Outside on the courtyard, the MV-22 was completely destroyed and the remaining pieces of the fuselage were burning ferociously. As the aircraft had dropped its ramp, a swarm of drones pounced from their staged locations around the embassy and released a barrage of bombs directly over the Osprey. The explosions ripped the aircraft apart causing it to spin, flip, disintegrate, and burst into flames. No one got out alive. The Cobras picked up the movement of the drones on their thermals, but there was nothing they could do, but watch. The entire series of events took only seconds and then the drones all darted off in different directions. The Marines never had a chance and didn't know what hit them. Nothing in their training had prepared them for this kind of attack and certainly none of their gear or weapons were set up to defend against it. Sadly, even though the military was well aware by this point that the compound was under a sustained attack by unmanned systems, they still failed to change their tactics and it cost a lot of young men their lives. It was a total loss of 14 Marines plus the crew. The aircraft alone cost the taxpayer over $120 million dollars as it was configured. The terrorists had written a new chapter in warfare. From here on, big expensive aircraft had just become obsolete. A dozen, small, inexpensive drones just executed a decisive, successful attack against some of the most advanced weapons in the US military's arsenal. Not even the terrorists could fully appreciate the revolution in warfare that had just occurred and where it would eventually head.

"Abort, abort, abort," was all that you could hear on the radio as frantic radio traffic filled the net. The rest of the MV-22 squadron waived off the landing site and turned back to sea in a holding pattern. "Mongol One, do you have eyes on the crash site?" Thunder's squadron commander radioed. "Thunder, this is Mongol One that's affirmative. No survivors. I am sorry." John radioed back as he watched the aircraft burn. "I told them. I told them. Damn it! I told them. If only I could have found a way to make management listen a year ago." John mumbled to himself as he continued to stare at the wreckage. Back in Washington, the drone feed left the room standing in stunned silence. Now Washington was frantically beginning to contemplate the fallout from their reluctance to allow preparatory strikes. Specifically, the White House refused to clear attacking a single target prior to the rescue and blame was already beginning to spread. From the looks of the military staff standing around the feed; they didn't have to say a word. Everyone knew whose fault it was. By the President hesitating to allow the Marines to take out the initial targets they had doomed the rescue mission. The President stood in expressionless silence as he watched the MV-22 burn. As everyone turned to the President, the stoic façade began to erode as he began to seethe with rage. There was no debate now. The President addressed the Secretary of Defense and ordered all known targets to be taken out. "Do it. Take out every damn location we know the drones have been flying from." "Yes, Mr. President."

11

Never Give Up

The failure of the rescue attempt had crushed moral. John could see the last bits of hope fading in the eyes of even the Marines. Looking at the group of people, John saw Caleb again crying and was momentarily overcome with the urge to choke the shit out of him as he thought of Caleb's previous threats to destroy Mark's career. John thought to himself as he stared with complete disdain at Caleb. "I wonder if that self-serving pajama boy is even aware that the same Mark he was going to report to the EEO officer and was already drafting nasty memos about to management is lying out there dead because he saved Caleb and everyone else huddled in the chancery? He is just the type that would say something crass to justify his perpetual cowardice like Mark volunteered for it and knew the risks." John then forced himself to return to the present predicament only after swearing he would hook up Stroh in the proper manner at a later date. John looked out at the group and spoke. "We aren't beaten. We are getting out of here." Pete Lowe spoke up from the crowd. "Great. How do you propose we accomplish that?"

John pulled the team together. "Matt, take the comms guy down to the sally port. I need you guys to fire up the suburban with the ECM package we never are allowed to use and tune it to whatever the hell frequencies the drones would operate on." "ECM" was the term for electronic-countermeasures. Over the years, nearly all government vehicles operating in high threat areas could be set up with sophisticated and effective jamming equipment to defeat improvised explosive devices. However, in typical fashion, State refused to allow the defensive technology to be deployed and used. Even stranger was the fact that not just the military, but the Secret Service, United Nations, and even NGOs had been smart enough to be running ECMs for years. Why State refused to use the important technology was beyond any rational man's comprehension and it was a miracle that no one to date had been killed as a result of not employing the technology. Then again, after seeing how State utterly ignored the UAS (unmanned aerial system) threat, it shouldn't be any surprise. John continued. "If you can't figure it out, then just go broad spectrum. Max the power and make sure the antenna jams up instead of out. My thoughts are that if we can push out enough power and jam the drones, we can create a bubble that will give us a shot at breaking out of the embassy to the alternate LZ and getting out of here." No thanks to DS, John had worked with and used ECMs quite a bit while in the military. Although the technology had improved, the same basic principles were in effect and he knew how to leverage it. "No problem, but how is everyone going to get to the LZ?" Matt asked. John thought for a second before responding. "Honestly, I am not quite sure. I envision some crazy, balls out escape attempt because we don't have enough vehicles to put everyone in. Folks will have to walk with the vehicle to the field I guess. If any of the armored ones are running, we can at least use them to transport the wounded. We can't shuttle because it would leave anyone at the LZ unprotected. We all go at the same time. Either way, just get on it and let me know as soon as you are set up."

Back in the air, the orders were clear. The Cobras were to use their FLIR (Forward Looking Infrared) to try and find the point of origin or "POO" of the drones. If they could locate a POO, they were to destroy it. Further, Washington had passed on at least a dozen targets to the AV-8Bs to engage based on the intelligence and targeting data the Air Force put together during the day. In the interim, the MV-22s were to circle above the city until the attack aircraft finished taking out their targets for a possible attempt to land at the secondary LZ. With clear orders and a thirst for revenge, the pilots began to blast apart one target after another. Generally speaking, the collateral damage, even in a pure combat zone would have made these strikes prohibitive, but tonight the gloves came off. If you were in a building that happened to be being used to launch and recover drones, it was getting bombed. The strikes were decimating entire sections of the city and fires burned out of control across entire districts as the final targets were checked off the list. If the Ambassador were alive, she would be losing her mind, but then, if she had made some decent decisions, no one would be in this predicament in the first place. The Ambassador paid the ultimate price for misjudging America's enemies. She insisted on a political course that had willfully ignored the intelligence on Tunisia and went down with her ship. Still though, the Tunisians needed to be informed that the US had initiated a defacto war on in their country. A terse call placed to the Tunisian Ambassador in Washington served as the notice. Naturally, the Tunisian Ambassador was furious and demanded the strikes be halted, but he was utterly stonewalled. The only qualifier the Tunisian Ambassador was given was that if the Tunisian military moved to intercept or attack any US aircraft or personnel over their country, they would also be destroyed. For a seasoned diplomat schooled in the arts of ambiguity and negotiation, this message was shockingly clear. Washington now gave zero fucks and you had better step out of its way.

Meanwhile, back in the chancery the preps were being completed. Jennifer had gotten the embassy staff down to the sally port and staged to move. "John, everyone is accounted for." Jennifer confirmed. "Wes, as soon as the Marines finish loading the last casualty in the FAV, let me know. I am running upstairs where I can get a radio call out to the aircraft." John said. After running with Dave back through the chancery, John grabbed the radio. Dave stopped John for a moment and spoke. "John, you are truly a crazy bastard. You know this right? Do you think this will honestly work?" John lowered the radio and replied. "Honestly Dave, I don't know. We are making this up as we go and I am really running out of options. By that I mean maybe one or two other ideas." Dave seemed to get interested because he wasn't keen on the current plan. "Oh yeah? What's Option B John?" John laughed, "Everyone run like hell in different directions. I figure they will get a bunch of us, but a few of us will survive to tell the story. Other than that, we have to sit here and hope the cavalry comes up with a better plan real soon. If we are still here by morning, we are going to be overrun and it's pretty clear Washington doesn't have any other tricks up its sleeve. To be frank, I never had much faith in our government ever saving us anyhow. Washington will always do what is politically expedient. We are sacrificial pawns in a much bigger game that's been going on since the beginning of time. We are too low on the food chain for them to care beyond how it will affect the next election. My guess is we are on our own buddy. The fact is we are low on everything to include shooters. Even if we had the weapons and ammo we would need people that knew how to use them to hold off another attack. We need to get out tonight or we are going to die here." Dave stared out the window for a few long moments and contemplated their situation. He then visibly came to peace with what had to be done and spoke without turning his gaze. "We have kids to go home to. We are getting the hell out of here tonight one way or another. Radio the plan."

"Thunder, Thunder, this is Mongol One. Over." "Roger Mongol One. Go ahead." "Thunder, we need you to be prepared to extract the entire embassy staff at 0215 from the secondary LZ. We are going to use the ECM system in one of our vehicles to create a bubble that should jam any drones operating close enough to hit us. Break ('break' is used in long message traffic to split the transmission making it harder for the enemy to triangulate). We are going to leave the system running on the LZ so don't expect us to have any comms as you come in. If it goes bad, a red flare will signal you to abort. Over." "Mongol One, copy all. Stand-by for confirmation. Out." John leaned forward pressing his forehead against the window. "I hope the hell they will approve the mission Dave."

While Dave and John discussed contingency plans as the waited, the MEU planning staff was in heated debate over the extract plan John proposed. Major Lee, speaking to the other officers stated in plain language, "There is no way in hell we can send another Osprey in there. Everyone knows that there is no way we took out all of the drones. If that ECM bubble isn't big enough or doesn't work, we will lose everyone." The intel officer parroted Major Lee's position. "Sir, based on the overhead feeds, drones are still active in the area. We have knocked out probably ten or eleven of their bases, but our estimates based on the numbers of drones we have seen today are in excess of thirty sites. I can't speak to the ECM system's capabilities, but I can tell you for sure there are drones still operational." Colonel Wyatt turned to the communications officer. "Captain, is what the guys on the ground with the MSD telling us possible? Is that ECM system capable of effectively jamming the drones?" The communications officer was sweating. He knew the mission and ultimately the lives of everyone hinged on his answer. The weight of the responsibility was crushing. He had volunteered to be in communications just to avoid such situations. Now everything was riding on his assessment. Theoretically, it was possible, but he had no way of knowing if the drones had been modified or hardened against jamming or if their system was even functioning properly and optimized to jam drone frequencies. "Captain?" Colonel Wyatt prodded him for the answer. "Sir, the system should be capable of creating a bubble, but there are a lot of variables that may neutralize its ability to shield everyone." Colonel Wyatt fired back. "Captain, I need a straight answer. Will the system work?" The communications officer had been sitting in the MEU ops center monitoring the situation since it began. He knew that this was the last best hope for the embassy. No one else was coming for them anytime soon. If he was wrong, this was the end of the lives of the folks in Tunis and the careers for everyone on ship. It was time to man up. Leadership was about responsibility and this was his moment. "Yes, it will work sir." Colonel Wyatt had already lost too many Marines tonight, but he had been around long enough to know this is a reality of combat. People die and things go wrong. You can't give up and you have to push on. The decision and responsibility were now solely on his shoulders. The colonel looked over all of the intelligence once more as the staff sat quietly ready for him to fire off more questions. Then, in his stereotypical fashion, Colonel Wyatt stood, walked slowly to the front of the ops center without saying a word and placed his glasses on the table. "Gentlemen, we are bringing the American Embassy staff home tonight. I want everything we can put in the air up. I want our own EW (electronic warfare) assets fully integrated in the mission. I don't care if you have to black out the entire damn city. If they make it to the LZ, I want you to level everything from danger close to two blocks out around them, before you land a single bird. If there is a drone close by, I want it fried, jammed, or blown to pieces. We have to throw everything we got at them. This is our last chance gentlemen. Dismissed."

"Mongol One this is Thunder. Over." "Go ahead." "Mission is a go. Extract from the secondary LZ at 0215." "Roger Thunder, I copy extract from the secondary LZ at 0215 is a go." "That's affirmative and make sure you all keep your heads down. We are going to light it up before we come in." "Copy all Thunder. Out." "Dave, let's move. We don't have much time to prep." John said as they grabbed their packs and headed back down to the sally port.

"So we are just going to walk out there? Just like that?" James asked with disbelief. "Well, we have two options. We can test it and risk another drone flying in while we are screwing around or just go for broke." John answered. "McEwan, we have one chance. If we tip them off, by time we move for real, they will have come up with a way to still hit us." Matt said. John stomped his foot as he nodded in agreement. "Then it's settled. We all move together. All or nothing."

12

Movement to Extract

The MSGs completed the grim work of loading all of the bodies they were able to recover from inside the chancery into the FAVs parked in the sally port. On top of the bodies, they placed the stretchers with the wounded. It was a hellish setup, but space was at a premium and options were to say, limited. In preparation for the escape, the staff had donned the oversized and ill-fitting generic body armor and helmets the embassy stocked for high threat situations. This gave everyone the look of reporters, which were infamous for their inability to properly wear a helmet and body armor. James began to laugh. "All they need is a big 'PRESS' stenciled across their chests and it would be a perfect fit." Looks aside, it did provide an additional layer of protection for when they moved. The ECM truck would be the first to roll out and then pause for a short time. If nothing happened, the rest would follow it out. This was about the best operational test of the system's effectiveness they could get before fully committing everyone to leaving the protection of the chancery. From the sally port, it was about a two minute walk to the LZ. Any longer and they risked the chance the enemy could stage an attack while everyone was massed in the open. Any less and they wouldn't be able to all make it there in time. So, once John gave the order to go, everyone had to move to meet the tight timeline.

"It's time." John said to everyone assembled. The group quieted as they turned out all lights before the door was opened. "Okay, open the bay door." John whispered. As the door slowly lifted the stifling heat and humidity of the night air rushed into the sally port. The air was thick with the smell of smoke. It would certainly be one of those smells you would never forget. The ECM vehicle began slowly rolling toward the exit with its lights blacked out. Outside the bay door, a yellowish halogen light still shined across the parking area running on backup power. The light appeared to have drawn every bug in the city. It looked as if it was snowing underneath it as the warm glow reflected off of thousands of tiny insects flying about. "Can't have that backlighting us," Dave said as he shot out the light with his suppressed rifle. For now, everything was at least quiet and he didn't want to literally shine a spotlight on their move. The ECM truck cautiously exited the sally port and sat idle about 25 yards away. Still nothing... It crept another 25 yards and again halted in the still dark night. Perhaps, the air attacks launched against the drone sites had been more effective than everyone assumed. "Are you ready to make the call?" Dave asked John. John gave the nod and the entire staff began to slowly walk out of the chancery escorted by the vehicles carrying the dead and wounded. The mob resembled a herd of zombies as they hesitantly stumbled about in the dark. "Is this an evacuation or a Walking Dead episode?"

James mumbled as he was taking up the rear. "Hurry the hell up. We only have a couple minutes to get to the LZ."

Once the group linked up with the ECM vehicle, the entire entourage began to make their way towards the LZ. In the daylight, the walk was simple. Just follow some pavers laid in a trail of gravel, cross through a grove of Date Palms, and then traverse some grass manicured like a golf course. Easy right? Wrong. The group became a gaggle within seconds as they tripped, cursed, stubbed toes, ran into trees, got feet under tires, and wandered off in random directions. "Seriously, how damn hard is it to walk a couple hundred yards over open, flat ground?" James asked. Nonetheless, the gaggle finally made it to the open area behind the chancery. Unfortunately, that was the easy part. They now could hear the buzzing sound of drones approaching. Matt was pouring with sweat and he wasn't alone. Heat and humidity had nothing to do with it. This was fear. As the buzzing got louder he looked at Dave and John. "Moment of truth gentlemen." James was no different except that he smelled like a brewery as he sweated out what must have been a liter of Johnnie Walker he had clandestinely consumed in his hotel room the night before doing who only knows what. "See if we can find them." John said. The operators split the sky into sectors and began scanning for drones. James caught sight of one just as it crested the top of the chancery. "Six o'clock!" James yelled. The team spun on the drone just in time to see it lose control, stabilize into a hover, and then fly back the way it came. No sooner did it fly off, two more drones flew in and repeated the same maneuver.

"Dude, I think it worked." The communications officer said in relief. The group continued to move forward as three more drones came into sight and also reacted the same way. "John, how far away do you think they are when they turn back?" Matt asked. "I can't say for sure, but it looks like about 75 yards, which is better than I thought we would get based on the power output of this ECM system. The system must really be cranking out some wattage and we are probably all sterile now. Either way, it is just far enough to keep us out of range of their bombs so let's keep moving."

Watching the procession exit the chancery was one of the sniper teams, which reported on their exact movements. They had held their fire in order to allow the group into the open so the drones could inflict maximum casualties. However, the drones were having trouble and losing control every time they flew in for a bombing run. At first the operators thought the drones were simply malfunctioning, but then the snipers noticed that everyone seemed to mass around one vehicle. Wondering why it was so special, they began to observe it more closely and could clearly see some unique antennas. They had seen these same antennas on MRAP (mine resistant ambush protected) armored vehicles the Americans used in Iraq and Syria to jam remote detonation of hidden explosives. No strangers to electronic countermeasures, they knew instantly what the Americans were up to and radioed the information to the drone operators to shift tactics as the group began to disappear behind a wall that was obscuring the snipers' field of view.

The group was now less than a hundred yards from the LZ walking across the immaculately manicured lawn when suddenly bombs started falling around the embassy compound. "Get down!" Someone yelled as the blasts erupted. Like before though, the drones still seemed to only get within about 75 yards. However, two of the diplomats in tow began to flip out believing they were covered in blood as they felt a wetness covering their arms and face. "James, check them over." Dave yelled back. It took James only a second to figure out what had happened. The bomb had cut a sprinkler hose and the group was showered with water. Under the circumstances, scared to death and in the dark, it was forgivable confusion. Still though, it was all too typical considering the personalities. "Their fine. Keep moving." James passed back to the team. "They are trying to smoke us out John." Dave concluded. "Get up and keep moving. Pick up the pace." John yelled to the group now frozen in fear. He knew they had passed the point of no return. It was no use. The embassy staff began to freak out. Caleb was the first to start the panic and began wildly flailing his arms. "Damn it. Stroh is losing it." James noted. Then Stroh's rant began. "We are dead. We are all going to die. Don't listen to these idiots. We have to go back to the chancery. We were safe until these idiots forced us out on this death march." "Someone shut him up." Dave snapped. "Seriously, how did I know this guy was going to wet his pants and have a meltdown?" Dave asked. The operators knew that fear was contagious and if Caleb caused the group to bolt, the rescue would turn into a massacre. "Stay together. You leave the bubble and your dead." James yelled. "We're at the LZ. Form up a defensive triangle with the vehicles, get in the middle, and hold here." John passed to the group. The vehicles then circled up as if watching an old Western film where the settlers circled the wagons in preparation for an Indian attack. They made it to the edge of the LZ and still had a minute to spare. Suddenly, it all went to shit. Enemy drones started to pepper their immediate area with bombs. "Shit, I knew this was too good to be true! How the hell are they hitting us John?" Dave yelled as he dove for cover. John was right there with him as they rolled under one of the FAVs. "I have no damn clue, but they are doing it." Coming up to a knee, John could see he was about to lose the embassy staff, which were now in full blown panic. "Get to cover. Get in the triangle or under the vehicles. Jump in a vehicle if you can fit. Otherwise stay low." Looking up with his goggles, John could see drones flying overhead, except this time they were far higher than before. "Dave, they are flying above our bubble and lobbing in bombs. Not as accurate, but pray they don't get lucky."

The blasts continued and were now coming dangerously close. No doubt the enemy drone pilots were dialing in their aim with each sortie. "How long before extract John?" Dave yelled. "Should be now." John replied as he scanned the sky for any signs of the Marine aircraft. James however, was fixated on Caleb. He was in full panic and was trying to pull a wounded lady out of the FAV so he could get in. James crawled over to him and grabbed him by his belt and slammed him sideways to the ground. "Stay the fuck down and shut the door!" James growled at Caleb. Caleb began to loudly protest that everyone needed to run back to the chancery and started to leave the limited safety of the interior of the triangle the armored vehicles formed. Others began to follow as Caleb now started a mad dash back the way they came. The MSD operators were screaming for the staff that bolted to get back to cover. "I am not fucking chasing them John. Those idiots are on their own!" Dave said with decided emphasis. John knew at any moment the extract was going to begin and if they weren't there, they jeopardized the rescue. Looking out across the field, close to a dozen staffers were now in a blind sprint back towards the chancery as bombs continued to rain down. "Damn it! I need to get them back right now." John thought.

Approximately 600 yards away, the sniper team was waiting for the opportunity to engage. However, they didn't have a good angle. The group was almost completely obscured by the back perimeter wall. The team had set up to engage targets immediately around the chancery, but not the field in the back corner of the embassy. All they could do was report on what they saw until about eight individuals came running into the open. Making a quick hold over adjustment, the shooter steadied his aim on the man in the lead. He was in a full sprint, but holding a steady pace and direction making him relatively easy to track as he increased pressure on the trigger. Blasts from bombs the drones released continued to burst around the individuals, but so far, nothing had hit the mark. It was up to him to kill them before they made it back to cover. The shooter exhaled, stopping at his natural respiratory pause, as he continued to track the target through his scope. He continued to steadily increase pressure on the trigger until he felt the weapon recoil. Reacquiring his aim, he steadied his aim on the next target and began his press.

John and James watched as Caleb and a locally employed staffer approached the edge of the field wondering how the hell they were going to get them. Just as Caleb turned onto the ramp, he was driven to the ground by a bullet that ripped through his body armor, his body, and back out through the other side of the armor. The incredibly lethal .338 Lapua round, which was originally designed for big game hunting, effectively evacuated his chest cavity. A second later the staffer's head blew apart in graphic detail. Even from the distance it looked as if a water melon had been dropped on the pavement from a tenth story window. Both men fell lifeless to the ground. James took stock of the situation and then with almost no emotion said, "Well, if someone was going to find the sniper this time, Stroh was a good choice." After a long pause as the men stared at the dead bodies he continued. "I can't say I found the man to my liking, but too bad for the guy that followed him." Occasionally, you run across a real piece of work during your day-to-day activities that is so obnoxious, you wouldn't care if something awful happened to them. Further, you wouldn't feel guilty in the least, like social norms would demand, in the event it did. Caleb was this guy and unlike in the movies where everyone cries in anguish, more than a few people quietly celebrated Caleb's demise. The world was better off without him. Thankfully, Caleb's death also had the immediate effect of correcting the exodus he started. As soon as the first two men were shot dead, the rest turned and ran back to the LZ. In death, Stroh finally proved useful. Problem solved.

"Okay, man, where the hell are the MV-22's?" Erik, who up until this point had been very quiet, asked in an unmistakable, stressed the fuck out manner. John didn't have an answer and pretended he didn't hear him. It didn't matter anyway. Either Marines from the MEU were going to make the TOT or they weren't. There wasn't a damn thing anyone of them could do now on the ground. Dave also gave John the "what the fuck" look. Another blast showered the top of the FAV with shrapnel and threw clods of dirt and grass all over the people hunkered down. "This is getting too damn close man. If they aren't here in another minute, we need to move to some better cover. You good with that?" Matt asked. John didn't answer; he looked at his watch as he heard the rumbling of jet engines high overhead. It read 0215 just before the watch screen went blank. It wasn't just John's watch; the vehicles and the entire city around the embassy literally went black. The only lights they could see were from fires. As silence settled over the area, drones began to drop out of the sky like dead birds. "What the hell just happened?" Dave asked. John placed his head in his hands in relief. "They just hit the area with an EMP (electromagnetic pulse). They just fried anything with a microchip in it. Our radios are dead, but so are the drones." Dave looked out into the dark. "Yeah, but what about the snipers?" That too got a response as if it was now God answering the team's requests as the buildings surrounding the embassy compound blew apart in a fantastic display of firepower. The Marines had brought in F/A-18 Super Hornets to assist with the mission. The Super Hornets were advanced jets configured for both fighter and attack roles. At least one was configured for electronic warfare and had jammed and blacked out the area around the embassy as the rest of the jets released precision guided bombs into the surrounding structures. The snipers as well as any drone in a four square block radius around the embassy were obliterated in the strike.

The shockwaves from the blasts had enough force to knock the wind out of everyone sitting on the LZ and left everyone's ears ringing. The bombs were dropped at the absolute minimum distance from the group. They were so close that they could have killed some of the embassy workers, but it was a calculated risk. You could hear the giant, lethal chunks of vehicles and buildings thrown up by the bomb blasts pelting the ground all around them. As the falling debris subsided, the beautiful sound of the MV-22's came into focus. Two aircraft flew past then circled back and stopped in a hover over the LZ. The air coming from their giant propellers created another brown out as they descended side by side into the big field. This time, even before they had completely landed, the planes disembarked half a platoon of Marines that set up a security perimeter and then linked up with the MSD team.

"Mongol One?" a young Marine Sergeant asked. John smiled for the first time that night. "Yes, I am the actual." The Marine took a second to stare at everyone. "Okay sir; let's get you all out of here." The wounded were loaded first and immediately flown out. The second, third, and fourth waves evacuated the rest of the survivors and the dead. The MSD operators were the last to board, but before they did, they took a team of Marines back out to recover Mark's body. They weren't leaving until the whole team came home.

The sun was just beginning to rise over Tunis as the MV-22 began lifting off with the MSD team and last of the Marines. A gentle, golden light streaked across the sky as the sun broke over the horizon warming the faces of the men as they stared back one last time. The smoke from the burning embassy swirled in the rotor wash as the craft powered up for takeoff. As the aircraft gained altitude, it transitioned to its airplane mode, tilted forward, and began to speed away. Looking down, John replayed the events of the attack in his head as they flew over the remnants of the embassy. Soon though, the burned out compound faded from view as they gained altitude and left the city behind. As they flew past the beach and into the open water of the Mediterranean Sea, John reached into his pack where he kept a picture of his family. "I told you I was coming home," he thought. They weren't going back to the ship, but rather, directly to the base at Sigonella for treatment and debriefing. John stared out the back of the MV-22 at the water, which was flying with its ramp partially down. His mind was trying to process the events of the last 24 hours. In that short period of time, their team had lost almost half of its members, the embassy was wiped out, and the city of Tunis was left in ruin. Dave, Matt, and James had the same blank stares. No one talked or discussed what had just happened. There would be plenty of time for that and mourning their losses later. As they flew on, the rhythm of the aircraft's propellers and the waves of the sea passing under the aircraft became hypnotic and took their effect. One by one, the surviving team members, coming down from adrenaline highs, drifted into deep sleep for the rest of the flight.

13

The War Comes Home

The command center was remarkably state-of-the-art, but well hidden inside a non-descript office building. It housed a highly secretive and specialized group known only as Unit-J. Secrecy within Unit-J was of the absolute highest priority and unlike in other units, being captured or exposed literally meant death. The obsession with secrecy was not for nothing. Every intelligence agency in the world was trying to collect on the organization as their list of highly successful operations continued to grow. Still though, no one had successfully penetrated the unit. In fact, Unit-J had become something of a myth in the intelligence world. After seven years, no one had been able to so much as confirm even the existence of this phantom unit. The inability to find the unit led many to doubt it was ever real, which also provided a convenient bureaucratic cover for intelligence failures. The US government led the list of countries neither confirming nor denying Unit-J's existence. Nonetheless, US intelligence seemed to have an endless supply of new theories to explain and attribute how, why, and who was carrying out successful covert actions around the globe using very sophisticated techniques, methods, and equipment. Remember, where the world of politics intersects with the world of intelligence, if it doesn't exist, you can't "fail" to infiltrate it. However, like a black hole, even if you can't see it, you know it exists by the forces it produces on things around it. Unit-J's existence may have been questionable, but the results of its actions were not in dispute. As such, many had quietly come to believe that it couldn't be found by even the CIA because it was a new arm of the CIA and being used to carry out its blackest of black programs.

Inside the command center, at least two dozen men wearing tailored business suits were busy at computer terminals relaying information and orders, analyzing intelligence, tracking aircraft and shipping, and coordinating logistics over high speed encrypted networks. There was even a special financial cell, which was monitoring world stock markets and conducting dark pool trades of all types through anonymous intermediaries, which then rapidly moved profits via cryptocurrencies to offshore accounts to fund operations. This elite command cell had been painstakingly hand selected from warzones to boardrooms from around the world by special recruiters. Once recruited, a new member was individually trained in secret on specialized tradecraft, given new identities, and provided a blank check to procure cutting edge technology and weapons. However, if the recruit failed, they were permanently eliminated. The risk of leaving a loose end was unacceptable. After completing training, the new operator would be sent out to independently lead cells in various countries around the world. The most successful of these leaders were brought back and groomed for command. This literally was the stuff Hollywood made spy movies about. Each man had the finest pedigree when it came to training, education, and experience. There were no "C" students allowed and one's political and religious loyalties were heavily scrutinized. For these men, Ivy League educations were common and nearly all had been covertly paid for by extremely wealthy, silent donors. Degrees in computer programming, robotics, electronics and chemical engineering, finance, and cyber-security were as coveted as an international special operations background. These men were not recruited from the best in the USA, but rather, the best in the world. Unit-J had a global mission and nationality was immaterial.

The commander of this special operations unit was no less qualified. He had been recruited by the CIA out of Columbia University in part for his near flawless fluency in Arabic, Farsi, and French. After his initial training, he had spent over two decades with the Clandestine Service eventually becoming one of the top case officers running the CIA's covert, lethal, kinetic operations against terrorists around the world. He was already a legend in the Special Activities Division, which carried out some of the CIA's most dangerous missions. He knew every trick and even invented some of his own. The man wrote the book on modern covert operations. Beyond that however, the man didn't exist. His true identity had long been buried by the US government and he now operated from the shadows using remote means of communication. The man was the closest thing to a living ghost.

Today was special though. The commander had emerged from the shadows to appear in the flesh before Unit-J. No introductions were necessary. He had personally signed off on every member of his team and knew their lives in intimate detail even though they knew next to nothing about his. When he arrived, he was dressed in a fine Italian wool suit with handmade leather shoes, and wore a Patek Philippe wrist watch. For those of you not ranking amongst the world's elite, Patek Philippe carries one of the larger selections of watches costing north of a few million dollars, caters to the billionaires of the world, and refers to a $30,000 Rolex as cheap. His protective security team was no less qualified or well dressed. The detail didn't fit the gorilla in a suit image, but rather, were indistinguishable from other wealthy businessmen with the small exception of the well concealed, but highly sophisticated weaponry and suite of electronic countermeasures they carried.

Operations ceased as the commander walked into the command center. The room became silent as everyone stood in respect waiting for his address. With a truly theatric elegance, he began his speech in perfect English. "God is great." The staff all repeated, "Allah Akbar." Then the commander continued. "Today, we have made history and regained land stolen from the caliphate. Congratulations on your decisive victory in Tunisia. The Crusaders have been thrown back into the sea with heavy losses. In the process, Tunis was devastated by the Crusader's jets, which killed hundreds of innocent Muslims. Their hubris has turned Tunisia's population against their King and America. Their population will rise up and stand with us from here forward. Our recruiters have already enlisted over 300 new recruits just this morning. Warfare my brothers, will never be the same, but our struggle will only continue. There is no doubt that our enemies will adapt and seek us out even harder. Nonetheless, our success has made headlines. Our brothers cheer us across the globe and are rising up. We have given them hope and the tools to attack the unbelievers in ways they cannot defend against. So stand ready. Tomorrow we begin our attack on Washington. God is great and may peace be upon you." Without another word, the commander turned and walked out of the building onto the busy city street and quickly disappeared into Detroit's financial district.

Epilogue

The tragedy of Benghazi was that it was both foreseeable and preventable. The failures that led up to the catastrophe went far beyond just the unforeseen consequences of honest human errors. Perhaps counterintuitively, the system worked as it was designed up to the senior echelon of the Department of State's organization. The intelligence was accurate, timely, and clear that an attack against Americans in Libya was coming. Diplomatic Security elements in Libya also sounded the alarm and requested additional support. The warning systems clearly worked as designed. As a result, the Department's leadership and even Congress were briefed on the growing threats by the appropriate organizations. This provided solid proof the processes and procedures to pass critical information to decision makers also worked as designed. However, the Department's leadership failed to take appropriate actions in light of that critical threat intelligence. The flaw in the system was the assumption that when informed of a critical threat to American lives, our leadership would take appropriate actions to protect those lives. Remember, the failure of the State Department to act wasn't the result of a poorly informed decision. This was clear cut negligence in light of the information. We can debate the reasons why no actions were taken and point fingers at things like partisan politics, but the fact remains that those in power knew the huge risk Americans were taking in Libya and did not take the adequate precautions. Specifically, security was not sufficiently bolstered in light of the threats. In fact, senior management overruled the repeated requests of security experts and blatantly denied additional security measures. Organizationally speaking, how to prevent another leadership failure poses a very difficult challenge, which must be explored and effectively answered.

The Department is again failing to adequately address a terrorist threat. However, this time the threat is from unmanned autonomous vehicles otherwise known as drones. This book was written as an attempt to test one possible remedy to the Department's failure to adequately identify and respond to the threat before it becomes an operational reality. If successful, this book's message will contribute to preventing another Benghazi by publicly sounding the alarm bells.

The foundational question one must ask in order to analyze this organizational conundrum is: "When leadership fails to act on critical threat information that puts lives at risk, for whatever reason, should the inaction be ignored and/or tolerated?" If the answer is no, then the question becomes how is one to effectively force change at the top? The bureaucratic mechanisms already in place have been proven not to work. Remember, these mechanisms of our bureaucracy are intrinsically designed to inform, but not dictate to leadership. Ethics also matter. Are those with such information being responsible stewards of our nation's security by sounding the alarm and speaking out? So again, how then can change be made if the system is designed to allow leadership the final veto? If the foundational question is answered yes, then how can a free nation tolerate a state level organization that is unaccountable for gross negligence and the lethal consequences of its actions or inactions? How must an employee with critical information stay silent and live with the guilt that they could have prevented the deaths of fellow Americans? Could it ever be responsible to the nation, even if it was considered ethical, within the framework of the organization, to stay silent? Is a decision to remain silent something that can be justified when viewed through a lens of human rights? These are tough questions, but do have a correct answer.

Morally and ethically, there can be only one defendable position even if it is not convenient or easy for an organization to accept. You do what is necessary to save lives. The organization serves the interests of Americans, not itself, special interests, or partisan politics. A "public" servant's job is not to protect leadership or the organization, but the American people. Not only can gross negligence by leadership not be ignored or tolerated, but it must be proactively addressed when lives are placed at risk. In order to force that change when institutional systems fail, one must seek extraorganizational approaches that overtly identify the problem to leadership and to the public. This requires immense courage and all too often, professional sacrifice. The object is to place leadership in a position of political checkmate where they must effectively act in a manner a reasonable person would deem appropriate to survive in office. This accountability through transparency makes it very difficult for a political leader to refuse to take action because ignorance or denial of the issue would become impossible. To aid in this effort, the government writ large, must do more to enable and protect such actions, which are essential to maintaining the healthy function and accountability of our government to its citizens.

Addressing opposition arguments to exposing negligence, many fear "rocking the boat." Those "public" servants would actually say yes to my foundational question. It is not uncommon to hear such a position justified by the supposition of "you don't know what you don't know." These "don't make waves" bureaucrats may also suggest that in a democratic society, accountability for its leaders can only be retroactive, not proactive, and best administered through the possible loss of position or criminal and/or civil liabilities for the actions or inactions of leaders. This rationale often materializes in partisan hypocrisy and infers a gross level of negligence, which may rise to the point of demonstrated cognitive dissonance. In short, this dangerous mindset is of one who believes our government's leadership is omnipotent and any dissent should therefore be silenced. It further implies no one outside of the senior leadership can possible know what's going on or have an original thought. To those Orwellian adherents to the Great Oz, one should ask a few questions. "Do you honestly think it is best to let someone get killed? Do you honestly think someone in management could possibly justify why they were allowing someone to be killed? What if that person was you or a family member or a friend? Respective of the Department of State, exactly what part of your charge as a diplomat concerns making decisions about when it is okay to allow Americans to be left for dead by management? We are not talking about the military or CIA. Respective of a real world, not a straw man scenario, how exactly would the US have been harmed by someone sounding the alarm if it had prevented the deaths in Benghazi? The truth is leadership in a free country cannot effectively operate in a vacuum or in complete secrecy. Equally true is the fact that leadership is not served well by echo chambers. They need to hear dissenting arguments and it is essential those with critical information are heard.

Contrary to the logic of maintaining silence because "you don't know what you don't know," there are many times "you do know what you do know" and senior management is the one that doesn't have a clue. Operationally, the person in the field is almost always the first to perceive a threat or problem. The person on the ground is exactly whom leadership relies on for their information. This means you must take action to save lives and a top down driven response to threats is organizationally backwards. There is no brilliant wizard behind the curtain with all the answers contrary to what many would like to believe. Yes, it is comforting to believe some all-knowing being exists at the top of the management structure, but it is truly just a comforting myth. To allow critical threat information to be ignored is dereliction of duty and is at the root of organizational tragedies. Curiously, we already institutionally recognize action is necessary, even at the junior level, but somehow can't extend the same logic to the senior level when it comes to threats. Can you imagine a Secret Service Agent being told to keep quiet when it was clear there was an imminent threat to a person or venue the agent was protecting? No, of course not; we expect even junior agents to recognize a threat and take immediate action. Yet, somehow, when it comes to the lives of an entire diplomatic facility being placed at risk, it appears that everyone believed they should keep quiet and wait for management to act.

How is it when more, not less, lives are at risk, such as with Benghazi, people become reluctant to run forward and sound the alarm? A few theories come to mind. It could be everyone just assumes that the higher up the problem goes, the more likely someone else is already dealing with it. It could also be the intangible "immediacy" of the threat. If there is any doubt an attack is in process, there is a chance it won't happen and that causes people to hesitate sounding an alarm. After all, how can you prove an attack that never happened was prevented? No one wants to look stupid. Another interesting fact is that during the investigations of previous terrorist attacks, it is well documented that many people had information that could have prevented the attack, but reasoned away what they saw or knew. They failed to speak up because they didn't want to seem alarmist or to be wrong. Conformity is undoubtedly a powerful motivator in these cases, which certainly has institutional implications for security. A final theory may be as simple as risk versus reward. The individual knows that saving lives is the right thing to do. However, the difference is how directly it involves embarrassment to senior leadership. There is real fear of retaliation by senior leadership for sounding the alarm that draws attention to their failure(s), but no fear or danger in remaining quiet. Therefore, doing nothing is the safer choice. Considering all of the above, in practice, the answer is probably a little of all of those factors play into one's decision on whether to act or not. However, the discussion in many respects is purely academic because it shouldn't matter. If there is a threat to life, none of those theories for inaction are justifiable even if they are understandable.

It is important to still note that more often than not, lack of action is because leadership is not informed. Remember threat reporting is driven from the bottom up. Internal mechanisms are generally sufficient to address the most common shortfalls and should always be exhausted first. However, in the event senior leadership is fully aware of an impending threat to life, but fails to act, or senior leadership is unaware of a threat and has been bureaucratically firewalled from that critical information, then one must initiate special measures to prevent the loss of life. Morality, unlike ethics, is not negotiable. Sounding the alarm to save lives is for good. The only things hurt by sounding the alarm are the egos and careers of bad leaders and managers. Not only is this warranted, but both morally and ethically defendable. Bad leaders and managers do not deserve to hold their office. As such, if leadership fails to take action, is there a specific mechanism that could be used to prod those in power to take action in a timely manner? The truth is that there are many possible routes one could take. You never know if one route will work, but you must try and try again. This book is just one such extraorganizational attempt to address the growing threats from unmanned autonomous technology. It was never intended to be a literary masterpiece. Rather, it is a real-world institutional experiment with profoundly serious implications. Only in time will it be seen if the warnings imbedded in this book are heeded. If nothing else, I hope that I performed my duty to sound the alarms to the best of my ability.

Like the terrorist threats to the American diplomats in Libya in 2012, the threats from autonomous unmanned vehicles are real and growing today. It is not a future threat. It is a current threat. This book outlines a fictitious attack on an American diplomatic facility, but is based on the very real capabilities of weaponized drones. A coordinated attack using drone swarms against Americans both overseas and domestically will happen. Let me repeat that assessment in different terms. Drones will be used in the very near future to kill Americans. The US Department of State has ignored years of warnings and now has been caught off guard again through negligence, not ignorance of the threat. The Department is now scrambling to mitigate the threats, but have ignored them so long it is currently woefully unprepared for such an attack. This is organizational failure and should not have been allowed to happen. If the Department had properly assessed emerging threats in a timely fashion and dedicated the appropriate resources, this would all be a moot point. This did not occur and as a result the development and fielding of countermeasures have not kept pace with the threat. People are being killed all over the world by unmanned vehicles and if immediate actions are not taken by the US government, the next could very well be another ambassador. This will require a massive dedication of resources, but the choice isn't optional. We have to solve this problem right now.

For decades, the US has enjoyed a near monopoly on remotely killing people, but those days are over. ISIS, in particular, has already demonstrated its ability to launch attacks with both unmanned aerial and ground vehicles and have killed many people in Iraq and Syria. In fact, ISIS has released numerous videos showing the destruction of armored vehicles, boats, and checkpoints using drone delivered munitions. Further, ISIS has also developed and used remotely controlled car bombs to precisely attack military bases and troop defenses. What do you think will happen when they get their first self-driving Tesla or Amazon delivery drone? With increasing tensions between the US and nation states such as Iran, Turkey, Syria, China, North Korea, Venezuela, Mexico, and Russia, it won't be long before state level entities are passing their much deadlier drone technology to their proxies just as they do with nearly every other weapon system they develop. Disturbingly, this technology is far cheaper and easier to proliferate than other weapons suggesting that we will witness a very rapid increase in drone use in the coming months and years. Hezbollah is already believed to have been using Iranian drone technology to successfully attack Israeli targets both on land and sea. In the Ukraine, rebels are believed to have used a simple drone to attack a weapons depot and destroy over a billion dollars' worth of munitions in a single operation costing as little as a few thousand dollars. Even Mexican cartels are actively using drones to conduct reconnaissance and surveillance of the US southern border and to carry drug shipments. The threat posed by unmanned systems is difficult to understate when coupled with the resources of a state like China. It is revolutionizing warfare far faster and more drastically than anything we have seen in our lifetimes. Drones and their underlying autonomous technology will become scarier than atomic bombs within the next decade. I mean this literally. Drones are a far scarier threat than atomic bombs because of their precision, anonymity, reach, effectiveness, persistence, ability to proliferate, and their low cost. All of these factors mean that they will be regularly used unlike nuclear weapons. Their ability to penetrate beyond defenses and carry out pinpoint assassinations with impunity from anywhere in the world is only just the beginning and has already become the state's choice tool for assassination. Our policy makers are simply not prepared for what's coming next and America will be caught off guard once again if this warning goes unanswered.

With state level support, the attack this book foretells is not just possible, but is likely being planned somewhere in the world right now. Yesterday was the time to deploy tiered defenses against unmanned systems. "Now" is already too late. These countermeasures must be robust, redundant, persistent, mobile, and globally deployed or the weaknesses will be quickly identified and exploited. "A fix in the works" is not acceptable at this point. Institutionally, those leaders charged with developing countermeasures failed to proactively identify and counter this threat and should be replaced. Further, those responsible for forecasting future threats to our diplomatic facilities must also be brought to task. Drones are not some futuristic nonsense, but the reality of all current battlefields. They are killing people today! Let me be clear. Walls, barriers, and manned guard towers are no more of an obstacle to drones than medieval castles are today to jet bombers. If you are a VIP, you need to internalize the fact that you are the primary target and men with earpieces will not be able to protect you from this threat without the proper technology. Borders, customs inspections, biometrics, and passports are a joke for a drone operated from a different continent, which flies or even sails underwater unseen to its target. Drones also don't care if they carry a dirty bomb, Anthrax spores, or VX nerve agent over a city and far above any of the billions of dollars' worth of early detection equipment deployed around the US and world. Machine guns, rockets, and missiles are of no use because they cannot effectively hit or even locate small drones flying amongst crowds through city streets. Large, manned aircraft are also nearly useless and too costly to employ against cheaply produced swarms of highly lethal drones. Even the big armored vehicles and tanks will be proven to be obsolete against smaller, faster, and more agile drones just as the big battleships of World War II were sunk by swarms of carrier launched aircraft. The future is now. We must proactively move ahead of the threat or we will suffer greatly. Our weapons, both offensive and defensive are obsolete. A new era of warfare has dawned. This is your warning.

The Next Benghazi

www.blacksmithpublishing.com

www.ingramcontent.com/pod-product-compliance
Lightning Source LLC
Chambersburg PA
CBHW022057020426

42335CB00012B/723